AF361667

Radiation Sensing

Radiation Sensing: Design and Deployment of Sensors and Detectors

Editors

Kelum A. A. Gamage
C. James Taylor

MDPI • Basel • Beijing • Wuhan • Barcelona • Belgrade • Manchester • Tokyo • Cluj • Tianjin

Editors
Kelum A. A. Gamage
University of Glasgow
UK

C. James Taylor
Lancaster University
UK

Editorial Office
MDPI
St. Alban-Anlage 66
4052 Basel, Switzerland

This is a reprint of articles from the Special Issue published online in the open access journal *Sensors* (ISSN 1424-8220) (available at: www.mdpi.com/journal/sensors/special_issues/radiat_sens).

For citation purposes, cite each article independently as indicated on the article page online and as indicated below:

LastName, A.A.; LastName, B.B.; LastName, C.C. Article Title. *Journal Name* **Year**, *Volume Number*, Page Range.

ISBN 978-3-0365-1440-6 (Hbk)
ISBN 978-3-0365-1439-0 (PDF)

Contents

About the Editors

Kelum A. A. Gamage is an associate professor at the James Watt School of Engineering and a co-director of the Center for Educational Development and Innovation at the University of Glasgow, UK. Kelum has authored over 100 peer-reviewed technical articles, including book chapters, and holds a patent for a system designed to image fast neutron-emitting contamination (patent No. GB2484315, April 2012). His research group activities are primarily focused on the development of novel radiation instrumentation and techniques for nuclear decommissioning and security applications. He is an editor for *Sensors* (publisher: MDPI, Switzerland, ISSN 1424-8220), *Radiation Protection and Dosimetry Journal* (publisher: Oxford University Press) and *Education Sciences* (publisher: MDPI, Switzerland, ISSN 2227-7102). Kelum is a chartered engineer of the Engineering Council (UK), a senior fellow of the Higher Education Academy, a fellow of the IET, a fellow of the Royal Society of Arts and a senior member of IEEE.

C. James Taylor holds a personal chair in control engineering at Lancaster University, UK. His interdisciplinary research on statistical modelling and control of uncertain systems in the natural sciences and engineering involves applications spanning robotics, energy, health and the environment. He supervises projects in these areas for students across a spectrum of mechanical, electronic and chemical engineering disciplines. He is the author of over 150 peer-reviewed journal and international conference articles. His co-authored book *True Digital Control: Statistical Modelling and Non-Minimal State Space Design* was published by Wiley. Via the CAPTAIN software for MATLAB that he co-developed, his research has been transformed into readily accessible tools for data analysis across a wide range of industries. The toolbox is used worldwide for time-series analysis and control in disciplines as diverse as, e.g., hydrological modelling, electricity demand forecasting, etc.

Preface to "Radiation Sensing: Design and Deployment of Sensors and Detectors"

Radiation detection is important in many fields, and it poses significant challenges for instrument designers. Radiation detection instruments, particularly for nuclear decommissioning and security applications, are required to operate in unknown environments and should detect and characterise radiation fields in real time. This book covers both theory and practice, and it solicits recent advances in radiation detection, with a particular focus on radiation detection instrument design, real-time data processing, radiation simulation and experimental work, robot design, control systems, task planning and radiation shielding.

Kelum A. A. Gamage, C. James Taylor
Editors

Article

Integration of Ground- Penetrating Radar and Gamma-Ray Detectors for Nonintrusive Characterisation of Buried Radioactive Objects

Ikechukwu K. Ukaegbu [1,*], Kelum A. A. Gamage [2,] and Michael D. Aspinall [1]

[1] Engineering Department, Lancaster University, Lancaster LA1 4YW, UK; m.d.aspinall@lancaster.ac.uk
[2] School of Engineering, University of Glasgow, Glasgow G12 8QQ, UK; Kelum.Gamage@glasgow.ac.uk
* Correspondence: i.ukaegbu@lancaster.ac.uk

Received: 16 May 2019; Accepted: 16 June 2019; Published: 18 June 2019

Abstract: The characterisation of buried radioactive wastes is challenging because they are not readily accessible. Therefore, this study reports on the development of a method for integrating ground-penetrating radar (GPR) and gamma-ray detector measurements for nonintrusive characterisation of buried radioactive objects. The method makes use of the density relationship between soil permittivity models and the flux measured by gamma ray detectors to estimate the soil density, depth and radius of a disk-shaped buried radioactive object simultaneously. The method was validated using numerical simulations with experimentally-validated gamma-ray detector and GPR antenna models. The results showed that the method can simultaneously retrieve the soil density, depth and radius of disk-shaped radioactive objects buried in soil of varying conditions with a relative error of less than 10%. This result will enable the development of an integrated GPR and gamma ray detector tool for rapid characterisation of buried radioactive objects encountered during monitoring and decontamination of nuclear sites and facilities.

Keywords: ground-penetrating radar; gamma ray detector; sensor fusion; nuclear wastes; nuclear decommissioning; radiation detection; radiological characterisation

1. Introduction

The presence of radioactive objects in the shallow subsurface is a major public health risk because these objects can induce high levels of radiation above the ground. For example, a cobalt-60 source found buried at a depth of about 32 cm in a Cambodian hospital induced radiation levels of up to 60 mSv h^{-1} above the ground [1]. This is about 26,000-times the stipulated effective dose limit of 20 mSv per year [2]. Furthermore, chemical reactions in the soil can lead to the dissolution of these objects and subsequent contamination of groundwater. For example, the high energy penetrators used in ammunition are usually made from depleted uranium, which is a by-product of the nuclear fuel enrichment process. Many of these penetrators get lodged in the ground during military operations and become potential sources of groundwater contamination because of their high solubility in sand and other volcanic rock [3]. Therefore, it is important to promptly detect, and safely dispose these objects.

The first stage in the disposal of these buried radioactive objects is their characterisation. However, this process is challenging because of the difficulty in estimating the depth of these objects using traditional intrusive methods such as logging and core sampling [4,5]. Therefore, a number of nonintrusive depth estimation methods have been developed. These can be broadly divided into three categories, namely: empirical model methods; multiple photo peak methods; and shielding and collimator methods. The empirical model methods are based on establishing correlations between distinguishable features in part or all of the gamma spectrum and the depth of the buried radioisotope. They include: peak-to-valley ratio [6,7], peak-to-scatter ratio [8,9], principal component analysis [10–12], and machine learning [5,13,14] methods. However, these methods result in models whose parameters typically have no physical significance. Furthermore, the use of machine learning requires a significant amount of data for training. The multiple photo peak methods [15,16] exploit the difference in the attenuation of two energy peaks in the gamma spectrum in order to estimate the depth of the source. Consequently, they are limited to radioisotopes with two or more photo peaks that are sufficiently separated in the gamma spectrum.

The shielding and collimator methods [17–19] use different shielding and collimator configurations to obtain multiple measurements from which the depth of the radioactive source can be estimated. These methods have been shown to yield more accurate results compared to other methods [17] and can be used with any radioisotope. However, the required multiple measurements can only be acquired sequentially. This can significantly increase the data acquisition time because the acquisition of the spectrum of a buried source usually requires a long dwell time due to significant attenuation. In addition, in order to limit the minimum number of measurements required to estimate the depth to only two, the value of the bulk density of the soil is typically assumed to be known. However, the bulk density of soil depends on the current condition of the soil, and this varies from one location to another. Therefore, assuming a constant or generic value will result in errors in the estimated quantities. Furthermore, the use of historical values will not account for the changes in the soil density that would have occurred over time due to environmental factors such as rain fall and temperature changes.

Therefore, this work presents the development of a method for integrating gamma-ray detectors and ground-penetrating radar (GPR) for the retrieval of the soil density, depth and radius of a buried radioactive object. This eliminates the need for the soil density value to be known a priori. The method also used two horizontally-separated detectors to enable simultaneous acquisition of the required measurements, thereby solving the problem of sequential data acquisition. This will improve the rapid characterisation of buried radioactive wastes.

2. Theoretical Framework

For a radioactive point source buried in an air-soil half-space as shown in Figure 1, the flux F_p measured by the detector placed above the ground is given by [20]:

$$F_p = \frac{S_p A_r(E,\theta) C_e(E)}{4\pi \left(\frac{h+d}{\cos\theta}\right)^2} e^{-\mu_m(E)\rho_a \frac{h}{\cos\theta}} e^{-\mu_m(E)\rho_b \frac{d}{\cos\theta}} \tag{1}$$

where E is the energy of the point source (keV), θ is the angle of incidence of the source with the detector (radians), d is the depth of the source in the soil (cm), S_p is the activity of the source (Bq) and $A_r(E,\theta)$ is the angular response of the detector to a point source of energy E incident at angle θ. This is a dimensionless quantity and is obtained by measuring the response of the detector to a point source at angles varying from 0–$\pi/2$. This calibration should be done with the collimator in place if the detector is to be used with a collimator. $C_e(E)$ is the detector's centreline efficiency (cps cm^2 Bq^{-1}) and is calculated from the flux due to a source of known activity placed at a known distance z along the centerline, i.e.,:

$$C_e = \frac{F_p 4\pi z^2}{S_p} \tag{2}$$

where μ_m is the mass attenuation coefficient of the point source at energy E (cm^2 g^{-1}), ρ_a is the density of air (g cm^{-3}), h is the distance from the ground surface to the centre of the detector and ρ_b is the bulk density of soil (g cm^{-3}).

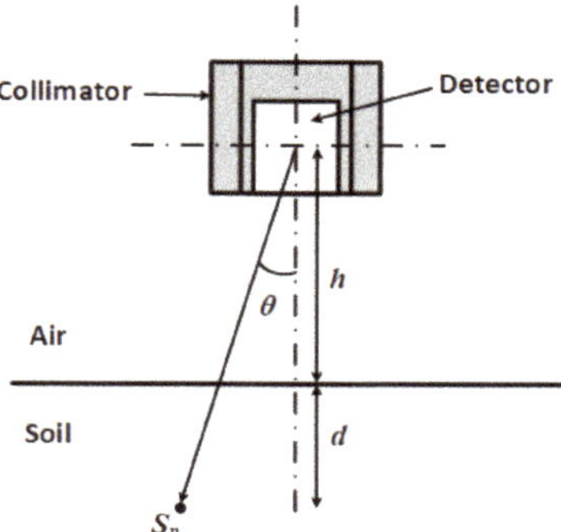

Figure 1. Geometry and parameters for estimating the flux (measured by the detector) due to the point source S_p in the soil.

If the buried object is assumed to be disk-shaped and the contamination is at most 1–2 mm below the object's surface, then it can be approximated as a planar disk source, and the flux F_a measured by the detector is obtained by integrating Equation (1) over the area of the disk, i.e.,:

$$F_a = \int_0^{2\pi} \int_0^r \frac{S_a A_r(E,\theta)C_e(E)}{4\pi \left(\frac{h+d}{\cos\theta}\right)^2} e^{-\mu_m(E)\rho_a \frac{h}{\cos\theta}} e^{-\mu_m(E)\rho_b \frac{d}{\cos\theta}} \, r\, dr d\phi \tag{3}$$

where r and ϕ are the radius (cm) and angle (radians) of the disk source in polar coordinates and S_a is activity per unit area (Bq cm^{-2}).

In most buried radioactive source surveys, the quantities of interest are the activity and depth of the source of the radiation; both of which are estimated from the ratio of two measurements [19]. In other words, the ratio of two measured fluxes F_1 and F_2 acquired using different detector configurations is a function that depends only on the source depth, i.e.,

$$\frac{F_2}{F_1} = ratio(d) \tag{4}$$

The depth estimated from Equation (4) can then be used to estimate the source activity using Equation (1) or (3) for a point or planar source. However, this two-measurement procedure assumes that the bulk density of the soil is known. This requirement can be eliminated by acquiring a third measurement [19]; however, this will increase the data acquisition time.

GPR has the potential of solving this density-dependency dilemma. A GPR system operates by sending electromagnetic signals into the ground and measuring any portion of the signal that is reflected by interfaces or objects in the signal propagation path. Using the illustration in Figure 2, the time t between the reception of the reflection from the ground and that from the disk source is given by:

$$t = \frac{2d}{v} = \frac{2d}{\frac{c}{\sqrt{\epsilon_b}}} \tag{5}$$

where v is the speed of the signal in the soil (m s^{-1}), c is the speed of light (299,792,458 m s^{-1}) and ϵ_b is the relative bulk permittivity of the soil (unitless). It should be noted that Equation (5) assumes that both the transmitting (Tx) and receiving (Rx) antennas are close to each other. Porous materials such as soil can be considered as a three-phase mixture of air, water and solid particles [21]. Therefore, their bulk permittivity is a function of the permittivities of these phases and their proportional composition in the material. Various formulas have been proposed to express this relationship; however, in a comparative study [22], it was shown that the formula based on the exponential mixing rule [21] with the exponent value of 0.65 gave the best result across a variety of materials. This formula is given by:

$$\epsilon_b^{0.65} = \left(\frac{\rho_b - W_c}{\rho_s}\right)\epsilon_s^{0.65} + \left(1 - \frac{\rho_b - W_c}{\rho_s} - W_c\right)\epsilon_a^{0.65} + W_c\epsilon_w^{0.65} \tag{6}$$

where the exponent value of 0.65 was obtained from the work of Dobson et al. [23], $\rho_s = 2.65$ g cm^{-3} is the solid particle density for soils, W_c is the volumetric water content (%), $\epsilon_s = 4.7$ is the solid particle relative permittivity for soils [23,24], $\epsilon_a = 1$ is the relative permittivity of air and ϵ_w is the relative permittivity of water, which is given by the real part of the modified Debye's equation [24], i.e.,

$$\epsilon_w = \epsilon_{w,\infty} + \frac{\epsilon_{w,0} - \epsilon_{w,\infty}}{1 + (2\pi f \tau_w)^2} \tag{7}$$

where $\epsilon_{w,\infty} = 4.9$ is the relative permittivity of water at infinity, $\epsilon_{w,0}$ is the static relative permittivity of water, f is the frequency of the GPR (Hz) and τ_w is the water relaxation time (s). Both $\epsilon_{w,0}$ and τ_w depend on temperature T (°C) and are given by Equations (8) and (9), respectively [25,26].

$$\epsilon_{w,0} = 88.045 - 0.4147 \times T + 6.295 \times 10^{-4} \times T^2 + 1.075 \times 10^{-5} \times T^3 \tag{8}$$

$$\tau_w = \frac{1}{2\pi}(1.1109 \times 10^{-10} - 3.824 \times 10^{-12} \times T + 6.938 \times 10^{-14} \times T^2 - 5.096 \times 10^{-16} \times T^3) \tag{9}$$

Combining Equations (5) and (6) will yield Equation (10), which can be solved simultaneously with Equation (4) to estimate both the soil bulk density and the depth of the source. This integration of the data from the GPR and gamma detectors can be considered as a type of low-level multisensor data fusion where data from different sensors are combined using physical models to enable or improve the estimation of physical parameters [27].

$$\left(\frac{2d}{ct}\right)^{1.3} = \left(\frac{\rho_b - W_c}{\rho_s}\right)\epsilon_s^{0.65} + \left(1 - \frac{\rho_b - W_c}{\rho_s} - W_c\right)\epsilon_a^{0.65} + W_c\epsilon_w^{0.65} \tag{10}$$

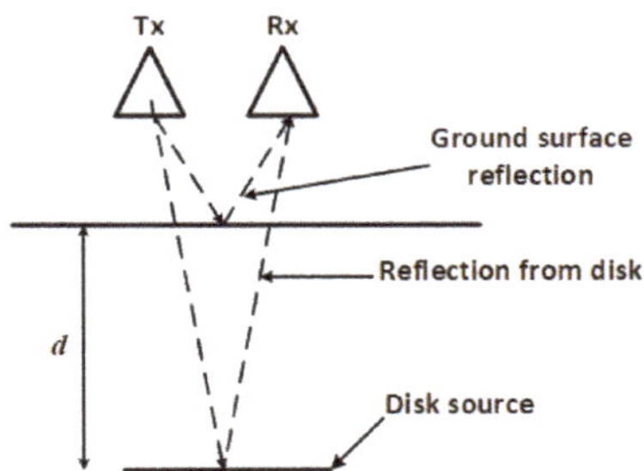

Figure 2. Operation of a ground-penetrating radar (GPR) system. Signals from the transmitter (Tx) are reflected by objects and detected by the receiver (Rx).

Another important consideration is how to arrange the sensors (i.e., the gamma detectors and GPR antenna) for efficient data acquisition. Preferably, the arrangement should be such that the sensors can operate simultaneously. Two ways of positioning two gamma detectors for the measurement of the radiation fluxes are illustrated in Figure 3. In the first arrangement, both detectors are vertically displaced by a fixed distance. However, this configuration makes it difficult to simultaneously measure the fluxes from both detectors because the field of view of the upper detector is completely or significantly occluded by the lower detector for small objects. This problem does not occur in the second arrangement where the second detector is horizontally displaced from the reference detector. This arrangement also has the additional advantage of allowing the GPR antenna to be mounted between both gamma detectors thereby creating a more compact sensor arrangement. However, the calculation of the angle of incidence (θ in Equation (3)) for the second detector needs to be modified to account for the horizontal separation. The modified expression is given by;

$$\theta = \arctan\left(\frac{a}{h+d}\right)$$
$$\text{where } a = \sqrt{(x + r\cos\phi)^2 + (r\sin\phi)^2} \tag{11}$$
$$\text{and } x \text{ is the horizontal separation.}$$

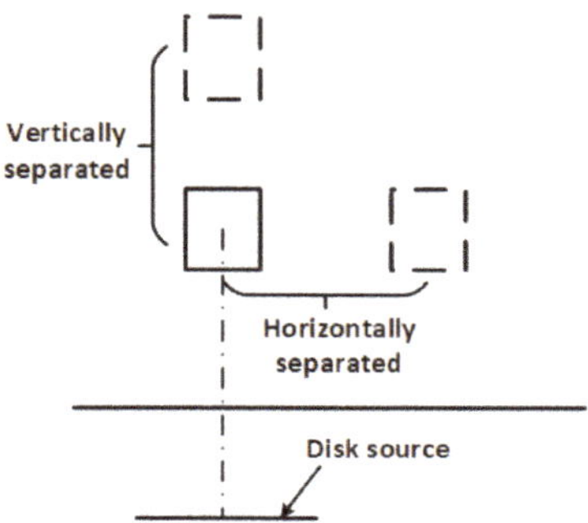

Figure 3. Two ways of arranging two detectors to measure the flux from the disk source. The horizontally-separated arrangement allows both fluxes to be measured simultaneously because none of the detectors is obstructed.

3. Materials and Methods

The numerical modelling and simulation tools used were Monte Carlo N-Particle Version 5 (MCNP5) [28] and gprMax Version 3.1.4 [29]. MCNP5 is a collection of software codes that is used to simulate the transportation of subatomic particles, e.g., gammas, neutrons, etc., and their interaction with materials using Monte Carlo statistical techniques. It is widely used in the modelling and analysis of nuclear radiation structures and systems and has been extensively proven to have good agreement with experimental results. gprMax is an open source software code used to simulate the propagation of GPR signals. At its core, gprMax is a finite-difference time-domain electromagnetic wave solver that uses Yee's algorithm to solve the three-dimensional Maxwell's equations. Its results have also been extensively validated with experiments [30].

3.1. Selection and Modelling of Sensors

The gamma detector used in the study was the CZT/500S from Ritec (Riga, Latvia). It is a hemispherical cadmium zinc telluride (CZT) semiconductor detector with a sensitive volume of 0.5 cm^3 (Figure 4a). The detector was chosen because of its size and good spectroscopic properties. In addition, unlike high purity germanium (HPGe) detectors, CZT detectors do not require a cooling system; therefore, they are very portable and easy to integrate with other systems. Figure 4b shows the simulated and experimental Cs-137 spectrum from the model and real detectors, respectively. A very good alignment of the spectrum key features can be observed. The tailing effect in the Compton valley of the spectrum from the experiment was due to incomplete charge collection caused by poor electron-hole mobility. This is a characteristic feature of CZT detectors. This feature was not modelled because of the additional complexity required. However, this will not affect the results of the study because the ratio of the area under the photo peak for two simulated spectra will be the same as that for two experimental spectra. The difference in the position of the Compton peak was likely due to nonlinearity in the real detector, while the higher background below 300 keV in the spectrum from the experiment can be attributed to backscatter from surrounding objects.

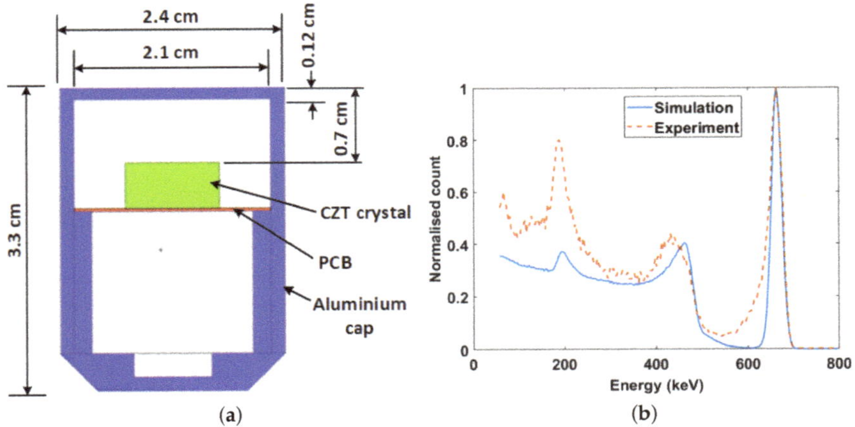

Figure 4. (a) MCNP5 model of the gamma detector. The crystal volume is 1 cm × 1 cm × 0.5 cm; (b) Experimental and simulated Cs-137 spectrum from the model and real detector.

The selected GPR antenna for the study was the 1.5-GHz antenna (Model 5100) from GSSI Inc. (Nashua, NH, USA). The gprMax model of this antenna is shown in Figure 5. The antenna consists of a pair of transmitter and receiver bow-tie antennas printed on a circuit board. The antennas are surrounded by microwave absorbers, which in turn are surrounded by a metallic shield. The entire assemble is enclosed in a polypropylene case. The development and experimental validation of the model can be found in [30,31]. It should be noted that the actual centre frequency of the antenna model was 1.71 GHz with a fractional bandwidth of 103%.

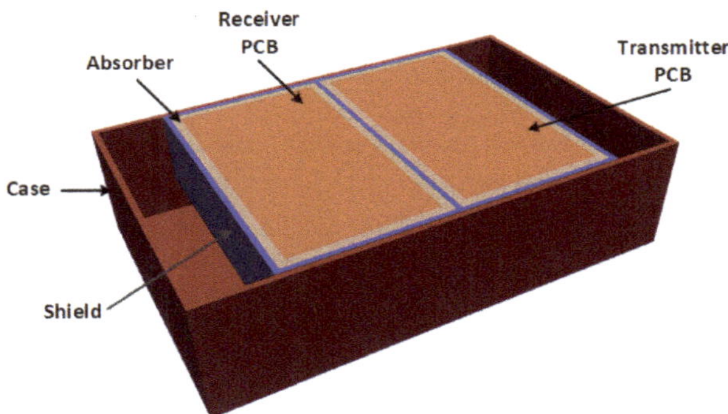

Figure 5. gprMax model of the 1.5-GHz antenna from GSSI Inc. The antenna dimensions are 17 cm × 10.8 cm × 4.3 cm (L×W×H). The skid plate underneath the casing has been removed to show the inside of the antenna.

3.2. Measurement Scenario Modelling

The measurement scenario was modelled both in MCNP5 and gprMax. The MCNP5 model of the measurement scenario is shown in Figure 6a. The radioactive object was modelled as a planar disk source with uniform activity. This is typical of stainless steel objects whose surfaces become activated by neutron flux in nuclear reactors [32]. The radioisotope used was Cs-137 with a photo peak energy of 662 keV. Each gamma-ray detector was placed in a cylindrical collimator with inner radius, thickness and height of 2.4 cm, 1.0 cm and 3.3 cm, respectively. The collimator was modelled as an alloy of tungsten (95% W, 3.5% Ni and 1.5% Fe) with a density of 18 g cm^{-3} [33]. The horizontal distance between the gamma detectors was selected such that it can fit the width of the GPR antenna. The antenna was modelled as a propylene box since it was not an active component in the MCNP5 simulation. The soil used in the model was a typical soil (51.4% O, 0.6% Na, 1.3% Mg, 6.8% Al, 27% Si, 1.4% K, 5.1% Ca, 0.5% Ti, 0.07% Mn and 5.6% Fe) with a dry density of 1.52 g cm^{-3} [34].

The gprMax model of the measurement scenario is shown in Figure 6b. This is a replication of the MCNP5 model using the gprMax antenna model described in Section 3.1. The detectors were modelled as metallic cylinders since only the lead collimator part of the gamma detectors will affect the GPR signals. The radioactive object was modelled as a metallic disk of thickness 0.5 cm. The two properties required to replicate the soil in gprMax were the bulk permittivity and the bulk conductivity. The bulk permittivity was calculated using Equations (6)–(9) at a temperature of 20 °C. The bulk conductivity was calculated using [35]:

$$\sigma_b = \frac{\sigma_w(\epsilon_b - 4.1)}{\epsilon_w} \tag{12}$$

where σ_b is the soil bulk conductivity (Sm^{-1}) and σ_w is the conductivity of pore water (0.05 Sm^{-1} [36]).

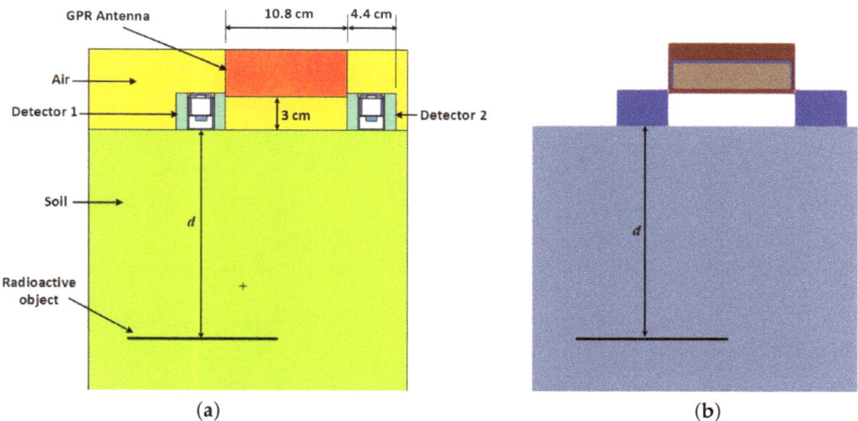

Figure 6. Model of the measurement scenario. The radioactive object is a metallic disk with Cs-137 radioactive contamination. (**a**) MCNP5 model of the measurement scenario. The gamma detectors are surrounded by 1 cm-thick lead collimators with an inner radius of 2.4 cm and height of 3.3 cm; (**b**) gprMax model of the measurement scenario. All labels and dimensions are the same as (**a**).

3.3. Simulation and Data Processing

Two sets of simulations were performed: MCNP5 simulations to measure the gamma fluxes due to the buried radioactive object and gprMax simulations to measure the time of flight (signal travel time) of the GPR signal to the buried radioactive object.

In the MCNP5 simulations, disk sources of radii of 3 cm, 9 cm and 15 cm were separately buried in the soil at depths varying from 12 cm–28 cm at 4-cm intervals. All the activities of the sources were normalised to 1 Bq cm^{-2}, unless otherwise stated. After simulation, a Gaussian function was fitted to the spectra from the gamma ray detectors in order to estimate the number of full energy photons detected. This is the required flux due to the buried radioactive object. The energy range used for the estimation was from 655–672 keV.

In the gprMax simulations, the radioactive object was also buried in the soil at depths varying from 12 cm–28 cm at 4-cm intervals. The GPR signal was then transmitted and the reflected signals recorded for processing. The first step in processing the GPR data was the subtraction of the antenna's system response from that acquired from the measurement scenario. The antenna's systems response is the measured response when the antenna is in air or free space. This subtraction process decoupled the reflection due to the ground surface from the direct signal from the transmitter to the receiver. This made the reflected signal from the ground surface easily identified. The required signal travel time was then the time between the ground reflection and the reflection due to the metallic disk. This process is illustrated in Figure 7a,b.

Using the estimated gamma fluxes and the signal travel times, Equations (4) and (10) were simultaneously solved to obtain the soil density, depth and radius of the buried radioactive object. These results are presented and discussed in the following section.

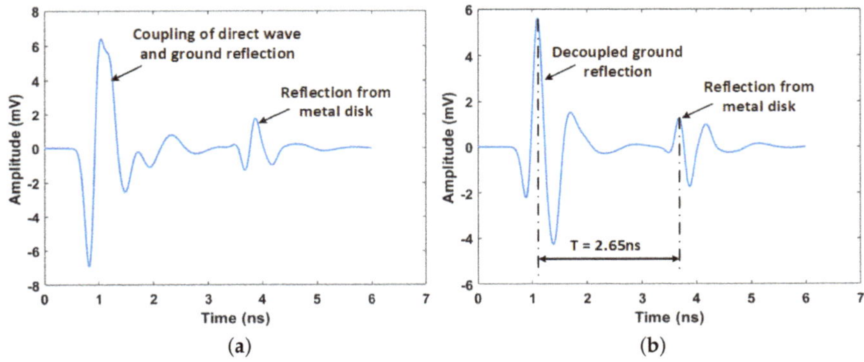

Figure 7. GPR signal for metal disk of a radius of 3 cm buried at 24 cm in dry soil, (**a**) Raw GPR signal with coupled direct wave and ground reflection; (**b**) GPR signal after subtraction of the GPR antenna's system response.

4. Results and Discussion

The calculated (solid lines) and simulated (markers) ratios of the fluxes (i.e., Equation (4)) from the gamma detectors for disk sources of different radii buried at different depths in the dry soil are shown in Figure 8. The uncertainty in the flux ratio was calculated using Equation (13), where δF_1 and δF_2 are the uncertainties in the fluxes from Detectors 1 and 2 as calculated by MCNP5. A decreasing dependency of the ratios on depth can be observed as the depth increased. This is indicated by the plateauing of the curves and the increasing error bars as the depth increased. This is caused by the exponential attenuation of the gamma rays as the depth of the source increased. This effect can be mitigated in practice by increasing the measurement time or by using a detector with higher efficiency. A decrease in the dependency of the ratios on depth can also be observed as the source radius increases. This is because the part of the source in the field of view of Detector 2 increases as its radius increases. Therefore, its measured flux will become increasingly the same as that measured by Detector 1 since the source has uniform activity.

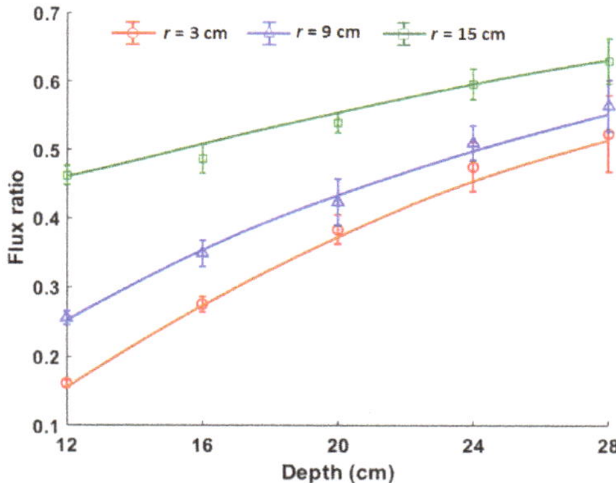

Figure 8. Flux ratio (i.e., F_2/F_1) for sources of radii of 3 cm, 9 cm and 15 cm buried at various depths in dry soil ($\rho_b = 1.52$ g cm^{-3}). The solid lines are calculated values, while the markers are the values from the simulation.

$$\text{Flux ratio uncertainty} = \left|\frac{F_2}{F_1}\right| \sqrt{\left(\frac{\delta F_2}{F_2}\right)^2 + \left(\frac{\delta F_1}{F_1}\right)^2} \tag{13}$$

The depths and densities obtained by simultaneously solving Equations (4) and (10) using the flux ratios in Figure 8 and the signal travel time from GPR measurements are shown in Table 1. The values in parentheses are the relative error in percentage. It can be observed that the estimated depths are within 5% of their actual values while most of the estimated densities are within 9% of their actual values. The density estimates with high errors are those obtained when the sources were buried at 12 cm. This is likely caused by the fact that the sources have a large incident angle with respect to Detector 2 when buried at shallow depths. This results in the reduction of the geometric efficiency of Detector 2.

Table 1. Simultaneously-estimated depths and soil densities for disk sources of different radii buried at different depths in dry soil. The values in parentheses are the relative error in percentage.

Actual Values		Estimated Values					
		$r = 3$ cm		$r = 9$ cm		$r = 15$ cm	
d (cm)	ρ_b (g cm^{-3})	d (cm)	ρ_b (g cm^{-3})	d (cm)	ρ_b (g cm^{-3})	d (cm)	ρ_b (g cm^{-3})
12	1.52	11.8 (2)	1.36 (11)	11.9 (1)	1.34 (12)	12.2 (1)	1.25 (18)
16	1.52	15.7 (2)	1.42 (7)	15.7 (2)	1.43 (6)	15.2 (5)	1.54 (1)
20	1.52	19.8 (1)	1.41 (7)	19.6 (2)	1.45 (5)	19.0 (5)	1.57 (3)
24	1.52	24.0 (0)	1.38 (9)	23.1 (4)	1.52 (0)	23.5 (2)	1.46 (4)
28	1.52	27.7 (1)	1.43 (6)	27.9 (0)	1.41 (7)	27.3 (2)	1.48 (3)

Table 2 shows the depth and density estimates for a disk source (3 Bq cm^{-2}) of a radius of 3 cm buried at a depth of 20 cm in soil of different densities and and volumetric water contents. The estimates in the first row were obtained using the proposed integrated GPR and gamma ray detectors approach. The values in the second row were obtained using the measurements from only the two gamma-ray detectors by minimising the following function:

$$\text{minimise: } \frac{(R_{calc} - R_{sim})^2}{R_{sim}} \tag{14}$$

where R_{calc} and R_{sim} are the calculated and simulated flux ratios respectively. It can be observed that the combination of the gamma detector and GPR measurements significantly improved the depth and density estimates especially at high densities and water contents. This is because the additional measurement from the GPR constrained the solution space to the correct values. The solution space can also be constrained by using a third gamma detector measurement; however, this will either increase the data acquisition time if the measurements are acquired sequentially or require the design of a complicated measurement geometry for simultaneous measurement of all three fluxes. Conversely, this GPR integration approach is fast, simple, and produces good results.

Table 2. Depth and density estimates for a disk source of radius 3 cm buried at a depth of 20 cm in three different soil conditions. The values in parentheses are the relative error in percentage.

Estimation Method	Soil 1 ($\rho_b = 1.67$ g cm^{-3}, $W_c = 15\%$)		Soil 2 ($\rho_b = 1.82$ g cm^{-3}, $W_c = 30\%$)		Soil 3 ($\rho_b = 1.97$ g cm^{-3}, $W_c = 45\%$)	
	d (cm)	ρ_b (g cm^{-3})	d (cm)	ρ_b (g cm^{-3})	d (cm)	ρ_b (g cm^{-3})
gamma detector and GPR	19.8 (1)	1.61 (4)	19.7 (2)	1.93 (6)	19.8 (1)	2.12 (8)
gamma detector only	19.17 (4)	1.48 (11)	17.6 (12)	1.5 (18)	16.83 (16)	1.5 (18)

Finally, the results presented so far assumed that the size (i.e., radius) of the disk source is known. However, this is typically not the case in practice. Therefore, the retrieval of the radius of the disk source was also investigated. Since this would require the estimation of three unknowns using two equations, the problem was reformulated as a constrained minimisation problem where Equations (10) and (7) are the objective and constraint functions, respectively. The result for disk sources of different radii buried in the soil at a depth of 12 cm is shown in Table 3. Good estimates can be observed as all of the estimated values had relative errors of less than 10% except the density and radius estimates for the disk source of radius of 3 cm. This large error in the estimates for the disk source of radius 3 cm is likely due to the large incident angle for Detector 2 at shallow depths, which reduced the number of gamma rays reaching the detector. This reduction in the flux measured by Detector 2 at shallow depths is more pronounced if the radius of the disk source is small. However, the results confirmed the ability of the integrated gamma detector and GPR method to estimate the key parameters of soil density, depth and radius of buried disk sources, simultaneously. Furthermore, this technique can also be used with other radioisotopes (e.g., Co-60) by substituting the mass attenuation coefficient at the photo peak energy of the radioisotope in Equation (3).

Table 3. Estimated depths, densities and radii values for disk sources of varying radii buried in the dry soil at a fixed depth of 12 cm. The values in parentheses are the relative error in percentage.

Actual Values			Estimated Values		
d (cm)	ρ_b (g cm^{-3})	r (cm)	d (cm)	ρ_b (g cm^{-3})	r (cm)
12	1.52	3	10.9 (9)	1.64 (8)	6.6 (120)
12	1.52	9	11.5 (4)	1.47 (3)	9.6 (7)
12	1.52	15	11.6 (3)	1.43 (6)	15.1 (1)

5. Conclusions

The integration of gamma detectors and GPR for nonintrusive characterisation of buried radioactive objects has been presented. The results showed that this integrated approach is able to retrieve the key parameters of soil density, depth and radius of disk-shaped radioactive objects buried in soil of varying conditions simultaneously. It also showed that by using two horizontally-separated gamma detectors, all the measurements required for the estimation process can be acquired simultaneously, thereby reducing the time associated with sequential data acquisition. However, the method is currently limited to objects having surface radioactive contamination that can be approximated by a disk. Therefore, there is a need to develop the method further to account for objects of different shapes. Finally, this study will form the basis for the development an integrated gamma detector and GPR system. Such a system will enable the rapid characterisation of buried wastes encountered during the decommissioning of nuclear sites and facilities.

Author Contributions: Conceptualization, I.K.U. and K.A.A.G.; methodology, I.K.U.; validation, I.K.U.; formal analysis, I.K.U.; investigation, I.K.U. and M.D.A.; writing, original draft preparation, I.K.U.; writing, review and editing, I.U., K.A.A.G. and M.D.A.; visualization, I.K.U.; supervision, K.A.A.G. and M.D.A.; funding acquisition, K.A.A.G.

Funding: This research was funded by the Engineering and Physical Sciences Research Council, U.K. (EP/N509231/1), and the Nuclear Decommissioning Authority, U.K. The APC was funded by Lancaster University.

Acknowledgments: The authors would like to thank Douglas Offin (National Nuclear Laboratory, U.K.) for industrial supervision of the project.

Conflicts of Interest: The authors declare no conflict of interest.

References

1. Popp, A.; Ardouin, C.; Alexander, M.; Blackley, R.; Murray, A. Improvement of a high risk category source buried in the grounds of a hospital in Cambodia. In Proceedings of the13th International Congress of the International Radiation Protection Association, Glasgow, UK, 13–18 May 2012; pp. 1–10.
2. IAEA. *Radiation Protection and Safety of Radiation Sources: International Basic Safety Standards*; Technical Report GSR Part 3; International Atomic Energy Agency: Vienna, Austria, 2014.
3. Bleise, A.; Danesi, P.R.; Burkart, W. Properties, use and health effects of depleted uranium. *J. Environ. Radioact.* **2003**, *64*, 93–112. [CrossRef]
4. Maeda, K.; Sasaki, S.; Kumai, M.; Sato, I.; Suto, M.; Ohsaka, M.; Goto, T.; Sakai, H.; Chigira, T.; Murata, H. Distribution of radioactive nuclides of boring core samples extracted from concrete structures of reactor buildings in the Fukushima Daiichi Nuclear Power Plant. *J. Nucl. Sci. Technol.* **2014**, *51*, 1006–1023. [CrossRef]
5. Varley, A.; Tyler, A.; Smith, L.; Dale, P. Development of a neural network approach to characterise226Ra contamination at legacy sites using gamma-ray spectra taken from boreholes. *J. Environ. Radioact.* **2015**, *140*, 130–140. [CrossRef] [PubMed]
6. Varley, A.; Tyler, A.; Dowdall, M.; Bondar, Y.; Zabrotski, V. An in situ method for the high resolution mapping of137Cs and estimation of vertical depth penetration in a highly contaminated environment. *Sci. Total Environ.* **2017**, *605-606*, 957–966. [CrossRef] [PubMed]
7. Varley, A.; Tyler, A.; Bondar, Y.; Hosseini, A.; Zabrotski, V.; Dowdall, M. Reconstructing the deposition environment and long-term fate of Chernobyl137Cs at the floodplain scale through mobile gamma spectrometry. *Environ. Pollut.* **2018**, *240*, 191–199. [CrossRef]
8. Adams, J.C.; Mellor, M.; Joyce, M.J. Depth determination of buried caesium-137 and cobalt-60 sources using scatter peak data. *IEEE Trans. Nucl. Sci.* **2010**, *57*, 2752–2757. [CrossRef]
9. Iwamoto, Y.; Kataoka, J.; Kishimoto, A.; Nishiyama, T.; Taya, T.; Okochi, H.; Ogata, H.; Yamamoto, S. Novel methods for estimating 3D distributions of radioactive isotopes in materials. *Nucl. Instrum. Methods Phys. Res. Sec. A* **2016**, *831*, 295–300. [CrossRef]
10. Adams, J.C.; Mellor, M.; Joyce, M.J. Determination of the depth of localized radioactive contamination by 137Cs and 60Co in sand with principal component analysis. *Environ. Sci. Technol.* **2011**, *45*, 8262–8267. [CrossRef]
11. Adams, J.C.; Joyce, M.J.; Mellor, M. Depth profiling 137Cs and 60Co non-intrusively for a suite of industrial shielding materials and at depths beyond 50 mm. *Appl. Radiat. Isot.* **2012**, *70*, 1150–1153. [CrossRef]
12. Adams, J.C.; Joyce, M.J.; Mellor, M. The advancement of a technique using principal component analysis for the non-intrusive depth profiling of radioactive contamination. *IEEE Trans. Nucl. Sci.* **2012**, *59*, 1448–1452. [CrossRef]
13. Varley, A.; Tyler, A.; Smith, L.; Dale, P.; Davies, M. Remediating radium contaminated legacy sites: Advances made through machine learning in routine monitoring of "hot" particles. *Sci. Total Environ.* **2015**, *521-522*, 270–279. [CrossRef] [PubMed]

14. Varley, A.; Tyler, A.; Smith, L.; Dale, P.; Davies, M. Mapping the spatial distribution and activity of 226Ra at legacy sites through Machine Learning interpretation of gamma-ray spectrometry data. *Sci. Total Environ.* **2016**, *545-546*, 654–661. [CrossRef] [PubMed]

15. Shippen, A.; Joyce, M.J. Profiling the depth of caesium-137 contamination in concrete via a relative linear attenuation model. *Appl. Radiat. Isot.* **2010**, *68*, 631–634. [CrossRef] [PubMed]

16. Haddad, K.; Al-Masri, M.S.; Doubal, A.W. Determination of 226Ra contamination depth in soil using the multiple photopeaks method. *J. Environ. Radioact.* **2014**, *128*, 33–37. [CrossRef] [PubMed]

17. Benke, R.R.; Kearfott, K.J. An improved in situ method for determining depth distributions of gamma-ray emitting radionuclides. *Nucl. Instrum. Methods in Phy. Res. Sect. A* **2001**, *463*, 393–412. [CrossRef]

18. Dewey, S.C.; Whetstone, Z.D.; Kearfott, K.J. A method for determining the analytical form of a radionuclide depth distribution using multiple gamma spectrometry measurements. *J. Environ. Radioact.* **2011**, *102*, 581–588. [CrossRef] [PubMed]

19. Whetstone, Z.D.; Dewey, S.C.; Kearfott, K.J. Simulation of a method for determining one-dimensional137Cs distribution using multiple gamma spectroscopic measurements with an adjustable cylindrical collimator and center shield. *Appl. Radiat. Isot.* **2011**, *69*, 790–802. [CrossRef]

20. Dewey, S.C.; Whetstone, Z.D.; Kearfott, K.J. A numerical method for the calibration of in situ gamma ray spectroscopy systems. *Health Phys.* **2010**, *98*, 657–671. [CrossRef]

21. Brovelli, A.; Cassiani, G. Effective permittivity of porous media: A critical analysis of the complex refractive index model. *Geophys. Prospect.* **2008**, *56*, 715–727. [CrossRef]

22. Ukaegbu, I.K.; Gamage, K.A.; Aspinall, M.D. Nonintrusive depth estimation of buried radioactive wastes using ground penetrating radar and a gamma ray detector. *Remote Sens.* **2019**, *11*, 7–14. [CrossRef]

23. Dobson, M.C.; Ulaby, F.T.; Hallikainen, M.T.; El-Rayes, M.A. Microwave Dielectric Behavior of Wet Soil-Part II: Dielectric Mixing Models. *IEEE Trans. Geosci. Remote Sens.* **1985**, *GE-23*, 35–46. [CrossRef]

24. Peplinski, N.R.; Ulaby, F.T.; Dobson, M.C. Dielectric Properties of Soils in the 0.3–1.3-GHz Range. *IEEE Trans. Geosci. Remote Sens.* **1995**, *33*, 803–807. [CrossRef]

25. Klein, L.; Swift, C. An improved model for the dielectric constant of sea water at microwave frequencies. *IEEE Trans. Antennas and Propag.* **1977**, *25*, 104–111. [CrossRef]

26. Stogryn, A. The Brightness Temperature of a Vertically Structured Medium. *Radio Sci.* **1970**, *5*, 1397–1406. [CrossRef]

27. Ukaegbu, I.K.; Gamage, K.A.A. Ground Penetrating Radar as a Contextual Sensor for Multi-Sensor Radiological Characterisation. *Sensors* **2017**, *17*, 790. [CrossRef] [PubMed]

28. Pelowitz, D.B. *MCNPX User's Manual: Version 2.7.0*; Los Alamos National Laboratory: Los Alamos, NM, USA, 2011.

29. Warren, C.; Giannopoulos, A.; Giannakis, I. gprMax: Open source software to simulate electromagnetic wave propagation for Ground Penetrating Radar. *Comput. Phys. Commun.* **2016**, *209*, 163–170. [CrossRef]

30. Warren, C.; Giannopoulos, A. Creating finite-difference time-domain models of commercial ground-penetrating radar antennas using Taguchi's optimization method. *Geophysics* **2011**, *76*, G37–G47. [CrossRef]

31. Giannakis, I.; Giannopoulos, A.; Warren, C. Realistic FDTD GPR Antenna Models Optimized Using a Novel Linear/Nonlinear Full-Waveform Inversion. *IEEE Trans. Geosci. Remote Sens.* **2019**, *57*, 1768–1778. [CrossRef]

32. Keith, C.; Selby, H.; Lee, A.; White, M.; Bandong, B.; Roberts, K.; Church, J. Activation product interpretation of structural material for fast critical assemblies. *Ann. Nucl. Energy* **2018**, *119*, 98–105. [CrossRef]

33. Gamage, K.A.A.; Joyce, M.J.; Taylor, G.C. A comparison of collimator geometries for imaging mixed radiation fields with fast liquid organic scintillators. In Proceedings of the 2011 2nd International Conference on Advancements in Nuclear Instrumentation, Measurement Methods and their Applications, Ghent, Belgium, 6–9 June 2011; pp. 1–5. [CrossRef]

34. McConn, R.; Gesh, C.J.; Pagh, R.; Rucker, R.A.; Williams, R. *Compendium of Material Composition Data for Radiation Transport Modelling*; Technical report; Pacific Northwest National Laboratory: Washington, DC USA, 2011.

35. Hilhorst, M.A. A Pore Water Conductivity Sensor. *Soil Sci. Soc. Am. J.* **2000**, *64*, 1922–1925. [CrossRef]
36. Ciampalini, A.; André, F.; Garfagnoli, F.; Grandjean, G.; Lambot, S.; Chiarantini, L.; Moretti, S. Improved estimation of soil clay content by the fusion of remote hyperspectral and proximal geophysical sensing. *J. Appl. Geophys.* **2015**, *116*, 135–145. [CrossRef]

Article

Reconstruction of Compton Edges in Plastic Gamma Spectra Using Deep Autoencoder

Byoungil Jeon [1,2], Youhan Lee [1], Myungkook Moon [3], Jongyul Kim [3] and Gyuseong Cho [2,*

[1] Intelligent Computing Laboratory, Korea Atomic Energy Research Institute, Daejeon 34507, Korea; bijeon@kaeri.re.kr (B.J.); youhanlee@kaeri.re.kr (Y.L.)
[2] Department of Nuclear and Quantum Engineering, Korea Advanced Institute of Science and Technology, Daejeon 34141, Korea
[3] Quantum Beam Science Division, Korea Atomic Energy Research Institute, Daejeon 34507, Korea; moonmk@kaeri.re.kr (M.M.); kjongyul@kaeri.re.kr (J.K.)
* Correspondence: gscho@kaist.ac.kr

Received: 22 April 2020; Accepted: 19 May 2020; Published: 20 May 2020

Abstract: Plastic scintillation detectors are widely utilized in radiation measurement because of their unique characteristics. However, they are generally used for counting applications because of the energy broadening effect and the absence of a photo peak in their spectra. To overcome their weaknesses, many studies on pseudo spectroscopy have been reported, but most of them have not been able to directly identify the energy of incident gamma rays. In this paper, we propose a method to reconstruct Compton edges in plastic gamma spectra using an artificial neural network for direct pseudo gamma spectroscopy. Spectra simulated using MCNP 6.2 software were used to generate training and validation sets. Our model was trained to reconstruct Compton edges in plastic gamma spectra. In addition, we aimed for our model to be capable of reconstructing Compton edges even for spectra having poor counting statistics by designing a dataset generation procedure. Minimum reconstructible counts for single isotopes were evaluated with metric of mean averaged percentage error as 650 for ^{60}Co, 2000 for ^{137}Cs, 3050 for ^{22}Na, and 3750 for ^{133}Ba. The performance of our model was verified using the simulated spectra measured by a PVT detector. Although our model was trained using simulation data only, it successfully reconstructed Compton edges even in measured gamma spectra with poor counting statistics.

Keywords: plastic gamma spectra; energy broadening correction; Compton edge reconstruction; deep learning; deep autoencoder

1. Introduction

Plastic scintillation detectors have poor spectroscopic characteristics because of poor energy resolution and absence of photo peak in the region of interest, which is above 100 keV. Therefore, it is hard to conduct radioisotope identification from plastic gamma spectra. Despite their weaknesses, plastic scintillation detectors have been widely used in radiation monitoring systems, e.g., radiation portal monitor, because they have unique characteristics such as low cost, are easily made in large volume, etc. Therefore, various spectral processing techniques have been reported for pseudo gamma spectroscopy of plastic scintillation detectors. Energy windowing [1–4], F-score analysis [5], energy weighted algorithms [6,7], and inverse matrix [8] are representative methods for pseudo gamma spectroscopy. However, these methods can be categorized as indirect pseudo gamma spectroscopic methods because it is impossible to directly identify the energy of incident gamma rays. Even though inverse matrix allows unfolding photo peaks in plastic gamma spectra, it works with spectra with good counting statistics only.

In contrast, there have been many studies on radioisotope identification, which is one of the purposes of gamma spectroscopy, using pattern recognition methods, such as library matching [9,10] and neural network-based classifiers [11,12]. Using library matching methods, it is possible to identify radioisotopes only if the library data are prepared to match with the measured data. In the case of neural network-based-classifiers, it is difficult to define practical accuracy. Although the outputs from neural networks are in the form of probabilities, they do not represent practical accuracy without confidence calibration [13].

In this paper, we propose a deep autoencoder model to correct the energy broadening effect, which is one of the weaknesses of the plastic gamma spectra. If the energy broadening effect is corrected, it is possible to conduct direct pseudo gamma spectroscopy differently from other methods because Compton edges are represented in gamma spectra. The datasets for this study were generated using the following procedure; establishment of probabilistic density function (PDF) library for radioisotopes using the results of Monte Carlo simulations, synthesis of PDFs with dependent random ratios for various combinations of radioisotopes, and generation of datasets via random sampling. For the generated and measured plastic gamma spectra, it has been verified that our model can reconstruct Compton edges from spectral measurement, even from spectra with low counting statistics.

2. Materials and Methods

2.1. Deep Autoencoder

An autoencoder is a type of an artificial neural network that generates an output signal whose dimension is identical to that of the input signal. Figure 1 shows a schematic of autoencoder architecture [14–16]. As shown in Figure 1, an autoencoder consists of two parts: encoder and decoder. In the encoder, inputs are encoded into internal representations with reduced dimensions in the latent space. In the decoder, internal representations are decoded into the reconstructed signal. In this unsupervised manner, the autoencoder is widely used for dimension reduction in many applications [17,18]. Furthermore, an autoencoder can be used for noise rejection. If we add noise signals to training data and train an autoencoder to reconstruct the input signal without the noise, the autoencoder is optimized to make a function to reject noise signals. A deep autoencoder is an autoencoder model whose encoder and decoder consist of three hidden layers or more [19].

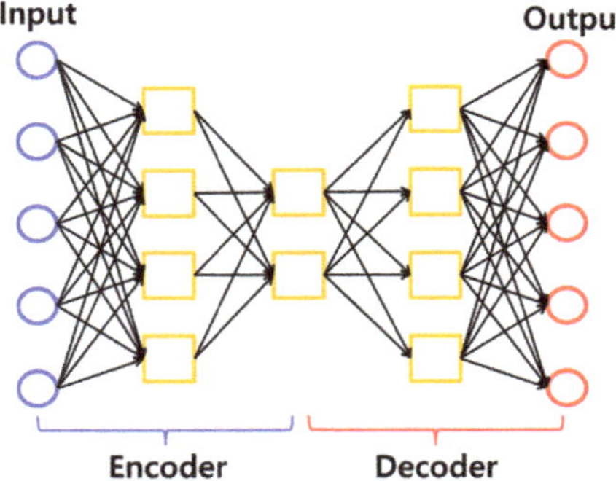

Figure 1. Schematic of autoencoder architecture.

2.2. Establishment of Datasets

2.2.1. Experimental Set-up

EJ-200 (cylindrical shape, dia. 30 × 50 mm, EJ technology) coupled with a PMT (R2228, HAMAMATSU) [20] and a preamp (E990-501, HAMAMATSU) [21] was used as a plastic scintillation detector. Optical grease (BC630, Saint-Gobain, Courbevoie, France) was applied at the junction between the crystal and PMT for optical coupling. For optical shielding, the crystal was wrapped with Teflon and black friction tape. A pulse processor (DP5G, Amptek, Hawthorne, NJ, USA) was used as a shaping amp with time constant of 2.2 µs and multichannel analyzer. A high-voltage supplier (NHQ 224M, ISEG, Lisboa, Portugal) was used to supply operating voltage to the detector. Experiments to measure gamma spectra were conducted in an aluminum dark box for the replenishment of optical shielding. The dark box consisted of a 10 mm thick aluminum case with an internal space of 440 × 440 × 899 (W × H × L) mm. The detector was placed on the shelf of the dark box, and the window of the detector was located at the center of the dark box. ^{22}Na, ^{60}Co, ^{133}Ba, and ^{137}Cs were used as gamma ray sources, and the position of the source was fixed at 5 cm from the detector window. Figure 2 shows our experimental setup. Energy calibration was conducted using a parametric optimization method [22].

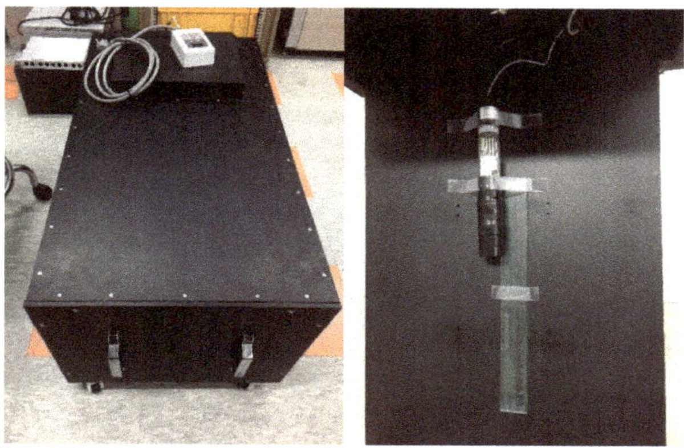

Figure 2. Aluminum dark box and experimental set-up.

2.2.2. Monte Carlo Simulation

To simulate plastic gamma spectra, we implemented a simulation geometry that was analogous to the experimental setup using the MCNP 6.2 software [23]. Compositions and densities of materials were defined from a material data report [24]. Gamma ray sources were defined as point sources. An F8 tally was used to simulate the spectral response of each source, and history number was set to 10^8. The F8 tally is also called a pulse height tally, and it is utilized when simulating deposited energy distribution according to energy bins, time bins, etc. Herein, we use the F8 tally with defining energy bins to simulate spectral response of our plastic scintillation detector. Energy bins for the F8 tally were defined as identical to energy calibrated channel bins. To acquire ideal and energy broadened pulse height distributions, F8 tallies were defined with and without a Gaussian energy broadening (GEB) card to acquire ideal and energy-broadened pulse height distributions, respectively. Coefficients "a", "b", and "c" for the GEB card were calculated by a parametric optimization method [22] using experimental spectra that were analogous to the measurement data to the maximum extent. Coefficient used for the GEB card is 0.006779 for "a", 0.3549 for "b", and −0.4999 for "c".

In MCNP 6.2, the energy broadening effect can be simulated with the use of a GEB option. When the GEB option is activated, all particle histories tallied in F8 tally are recorded after random sampling, which follows Gaussian probability distributions calculated by Equation (1):

$$f(E_0, a, b, c) = Ae^{-\left(\frac{2\sqrt{2\ln 2}(E-E_0)}{a+b\sqrt{E_0+cE_0^2}}\right)^2} \tag{1}$$

where A is a normalization constant; a, b, and c are GEB parameters; E is the broadened energy; and E_0 is the original energy before broadening.

2.2.3. Dataset Generation

The datasets were generated by random sampling and data synthesis using simulation data only. Before dataset generation, we prepared libraries of PDFs for ideal and GEB cases as follows. For pulse height spectra of ^{22}Na, ^{60}Co, ^{133}Ba, and ^{137}Cs simulated by MCNP code, each spectrum was divided by the integral value of itself for data normalization. With this procedure, each normalized spectrum could be represented as a PDF of detector response, because the summation of each normalized spectrum is one. After PDF libraries were created, we generated datasets as follows. First, ratios for PDF synthesis were selected as significant figures with first decimal place by dependent random sampling; the summation of synthesis ratios should be one to keep the synthesized results as PDFs. Some examples to explain the characteristics of dependent random ratios are as follows. If the synthesis ratio for ^{22}Na is one, the ratios for others should be zero. If the ratio for ^{22}Na is 0.1, the ratio for ^{60}Co is determined in the range of 0 to 0.9. If the ratio for ^{60}Co is determined as 0.5, the ratio for ^{133}Ba is selected in a range of 0 to 0.5. If the ratio of ^{133}Ba is 0, the ratio of ^{137}Cs is 0.4. With this spectral synthesis, data for multiple radiation sources in various ratios can be generated without additional simulation. Second, the number of samplings to simulate spectral data was then selected in the range of 40,000 to 100,000. By randomly selecting the sampling numbers, datasets with various levels of statistical uncertainties could be generated. This means that it is possible to build an autoencoder model with the ability to reconstruct Compton edges even from spectra with poor counting statistics with the generated datasets. Once the synthesis ratios and number of samplings were determined, PDFs were synthesized for ideal and GEB cases, and spectra were simulated via random sampling with the synthesized PDFs and the determined number of samplings. Next, spectra were normalized by total sum normalization, which can be represented as Equation (2),

$$x_{i,norm} = \frac{x_i}{\sum_{i=1}^{n} x_i} \tag{2}$$

where x_i is the ith element of the original data X, $x_{i,norm}$ is ith element of the normalized data X_{norm}, and n is the number of elements in the original data set.

In this manner, we established a procedure to generate datasets for the ideal case and GEB case paired with each other. Figure 3 illustrates the dataset generation procedure. With the established dataset generation procedure, we generated 60,000 spectra as a training set, 2000 spectra as a validation set, and 2000 spectra as a test set. Figure 4 shows the examples of the generated datasets.

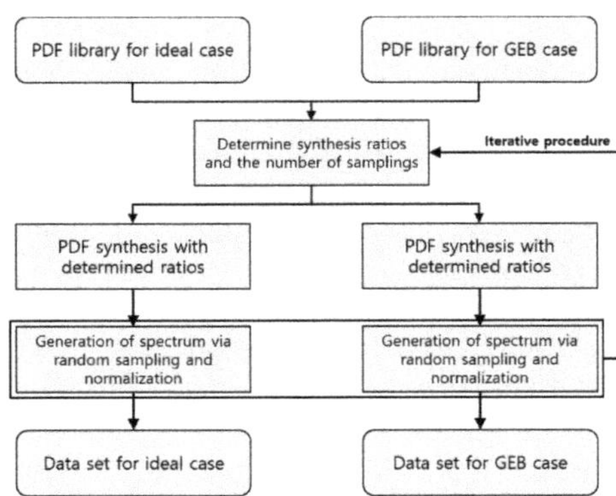

Figure 3. Flow chart of dataset generation.

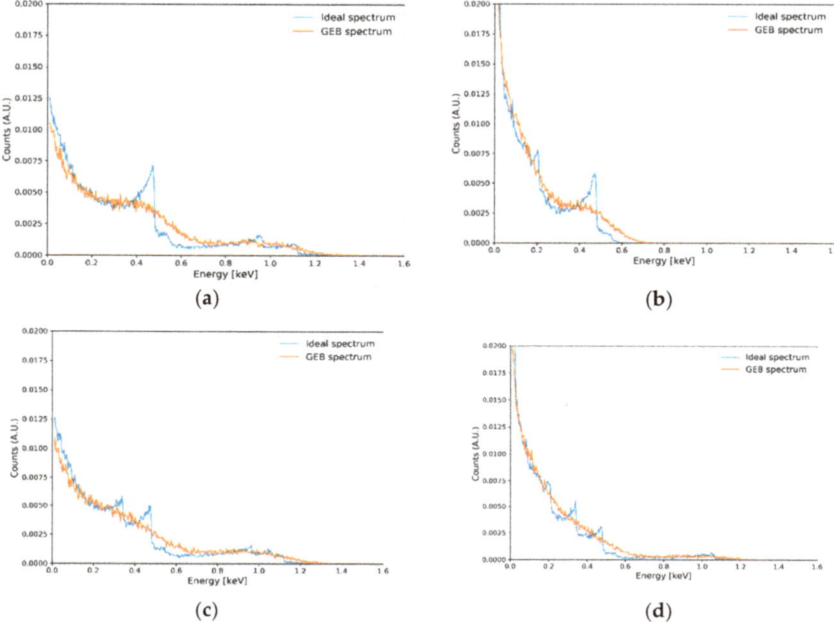

Figure 4. Examples of generated datasets for different synthesis ratios: (**a**) 70% ^{22}Na and 30% ^{60}Co; (**b**) 60% ^{137}Cs, and 40% ^{133}Ba; (**c**) 30% ^{22}Na, 30% ^{60}Co, and 40% ^{137}Cs; and (**d**) 40% ^{22}Na, 30% ^{133}Ba, and 30% ^{137}Cs.

3. Results

3.1. Results for Compton Edge Reconstruction with Test Set

The deep autoencoder was implemented in the Python environment using the Tensorflow [25] and KERAS [26] libraries. Hyperparameters for our autoencoder model were determined by trial and error as follows. The architecture of our model consists of three hidden layers as the encoder and three hidden layers as the decoder. The dimension of the input layer is 500, which means spectral data with 500 channel bins are provided as input to the autoencoder. The numbers of neurons in encoder layers are 200, 100, and 50, and the numbers of neurons in the decoder layers are 100, 200, and 500. This means that the input data are compressed by internal representations with dimension of 50 bins during the encoding process, and output with dimension of 500 bins is reconstructed from internal representations during the decoding process. For activation functions of hidden layers, a ReLU function was used for all layers of the encoder and the first and second layers of the decoder. For the third layer of the decoder, a sigmoid function was used as the activation function.

To train the deep autoencoder, training and validation sets for GEB case were given as input, and those for the ideal case were given as desired output. For data normalization, all data given to the deep autoencoder were presented as a response function in percentage units by dividing them into integral values of themselves and multiplying them by 100. In general, noise signals are added to the dataset with additional data processing procedure for an autoencoder to have the ability of noise reduction. In our problem, fluctuations in spectral data are coming from not noise signals but statistical uncertainties. By generating dataset via random sampling with randomly selected number of samplings, we can generate dataset with various level of counting statistics without additional procedure.

To compare reconstruction results with desired spectral data, mean absolute percentage error (MAPE) was used as a loss function, as described by Equation (3),

$$\text{MAPE} = \frac{100\%}{n} \sum_{i=1}^{n} \frac{O_i - I_i}{I_i} \tag{3}$$

where n is the number of channel bins, i indicates the ith channel bin, O is the Compton edge reconstructed spectrum, and I is the ideal spectrum given as desired output.

MAPE was employed for the following reason. Although there are various options for the loss function, most of them represent difference rather than relative difference between two data sets. Because the data used in this study are plastic gamma spectra, they have relatively high levels of counts in low and high channels. Therefore, other options are mostly affected by values in the low channel region, and values in the high channel region tend to be ignored. However, MAPE represents the relative difference between two data because the subtraction of two data is divided by one of them. Therefore, it can calculate the difference between two data with equivalent weights for the whole region of spectral data whether the level of count is high or low.

The deep autoencoder was trained with the ADADELTA optimizer [27] for established training and validation sets during 1000 epochs. Model checkpoint option was activated as a callback function to save the best model built during the training procedure by monitoring validation loss, and the best model in the training procedure was used as the final model. Figure 5 shows a schematic illustration of the training procedure of our model, and Figure 6 illustrates the training and validation losses during the training procedure.

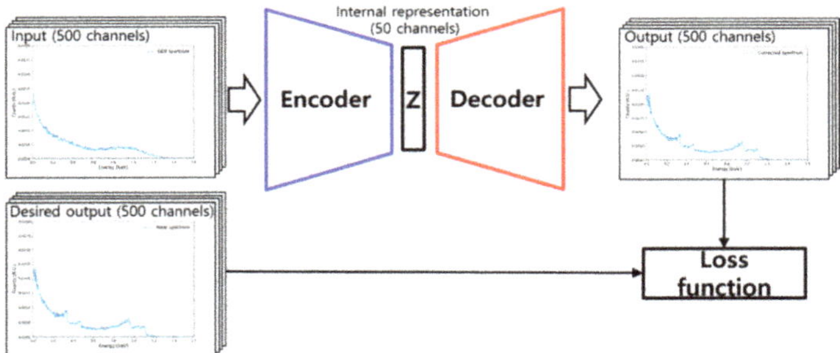

Figure 5. Training procedure of deep autoencoder.

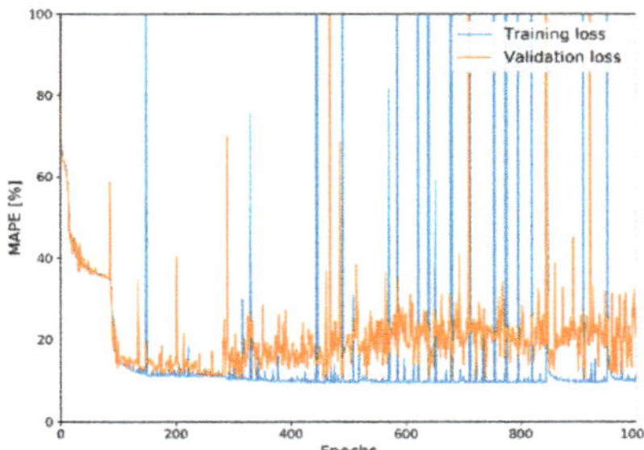

Figure 6. Historical plot of training and validation losses during training procedure. The best model during the epochs was stored with the use of the model checkpoint option and utilized as the final model.

The performance of Compton edge reconstruction for the trained deep autoencoder was tested using the generated test set. Averaged test loss was 20.019 for test sets. Figure 7 shows Compton edge reconstruction results for several spectra of single and multiple radioisotopes. The deep autoencoder reconstructed the Compton edges in plastic gamma spectra, even though the spectra contains statistical uncertainties. Information on spectra and their corresponding MAPE values are presented in Table 1. Synthesis ratios in Table 1 were not estimated by the deep autoencoder, but rather acquired during the test set generation procedure.

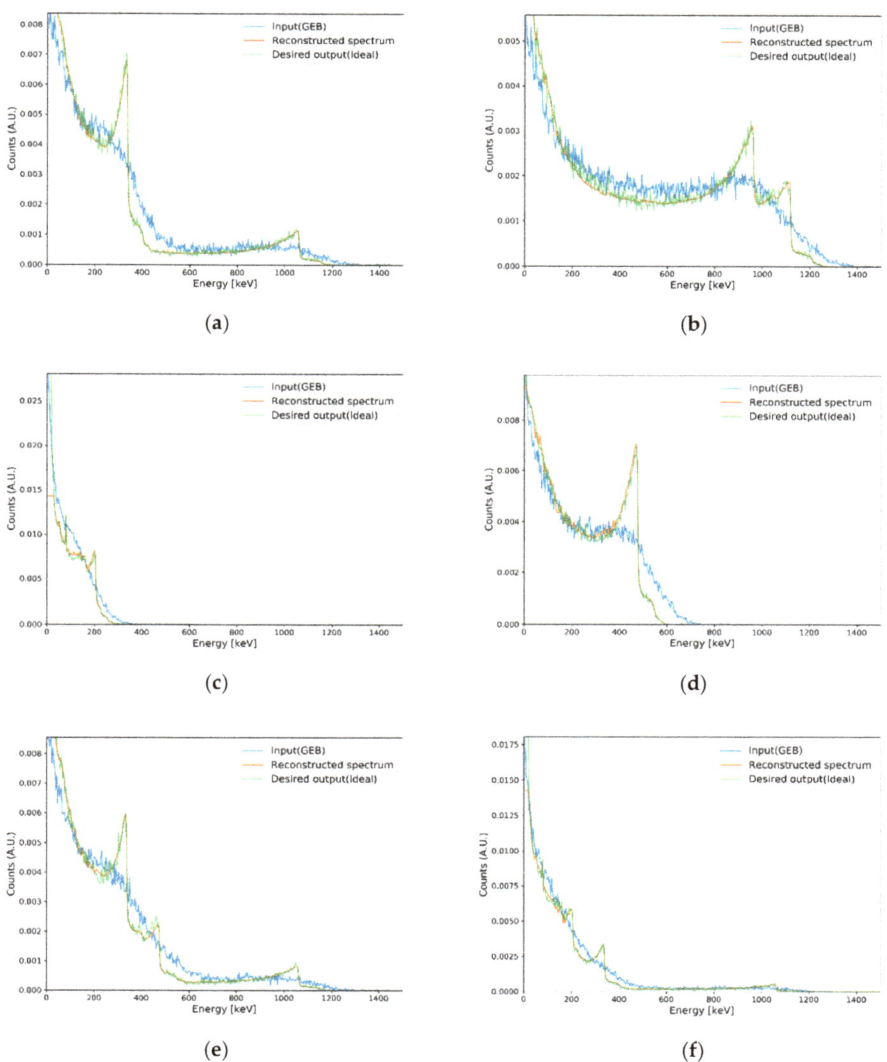

Figure 7. *Cont.*

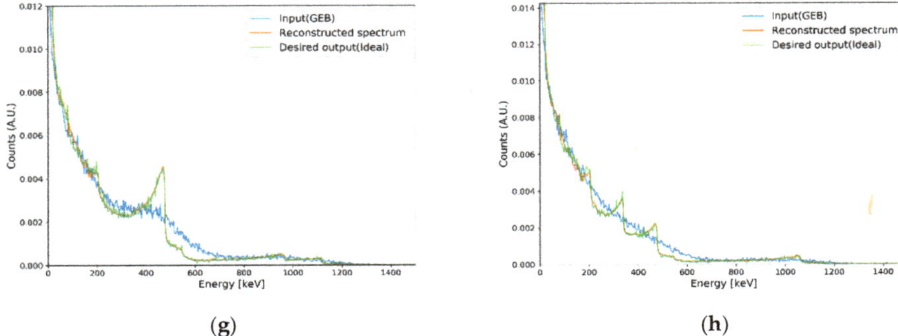

(g)   (h)

Figure 7. Results of Compton edge reconstruction for eight cases in the test set. Each synthesis ratio is (**a**) 100% ^{22}Na; (**b**) 100% ^{60}Co; (**c**) 100% ^{133}Ba; (**d**) 100% ^{137}Cs; (**e**) 70% ^{22}Na and 30% ^{137}Cs; (**f**) 50% ^{22}Na and 50% ^{133}Ba; (**g**) 20% ^{60}Co, 20% ^{133}Ba, and 60% ^{137}Cs; and (**h**) 40% ^{22}Na, 30% ^{133}Ba, and 30% ^{137}Cs.

Table 1. Information on seven cases in test set and their corresponding mean absolute percentage error (MAPE) values.

Case	The Number of Samplings	Synthesis Ratio				MAPE [%]
		γ_{Na}	γ_{Co}	γ_{Ba}	γ_{Cs}	
a	76,310	1.0	0.0	0.0	0.0	5.499
b	78,240	0.0	1.0	0.0	0.0	5.025
c	58,955	0.0	0.0	1.0	0.0	10.534
d	56,272	0.0	0.0	0.0	1.0	3.138
e	81,944	0.7	0.0	0.0	0.3	12.374
f	61,253	0.5	0.0	0.5	0.0	8.363
g	59,065	0.0	0.2	0.2	0.6	8.438
h	83,065	0.4	0.0	0.3	0.3	9.716

3.2. Results of Compton Edge Reconstruction for Experimental Data

Reconstructions of Compton edges using the trained deep autoencoder were also conducted for the experimental data. In the environment described in Section 2.2, plastic gamma spectra were measured from single to multiple radioisotopes with a measurement period of 3600 s. Background radiation was also measured, and background-subtracted measured spectra were provided as input data to our autoencoder. Figure 8 shows the results of Compton edge reconstruction for measured spectra of single and multiple radioisotopes. Compton edges marked in Figure 8 represent theoretical energies of each source calculated by the following equation [28] (p. 51),

$$E_{CE} = E\left(1 - \frac{1}{1 + \frac{2E}{m_e c^2}}\right) \tag{4}$$

where E is the energy of incident photon and $m_e c^2$ is the rest-mass energy of the electron (511 keV).

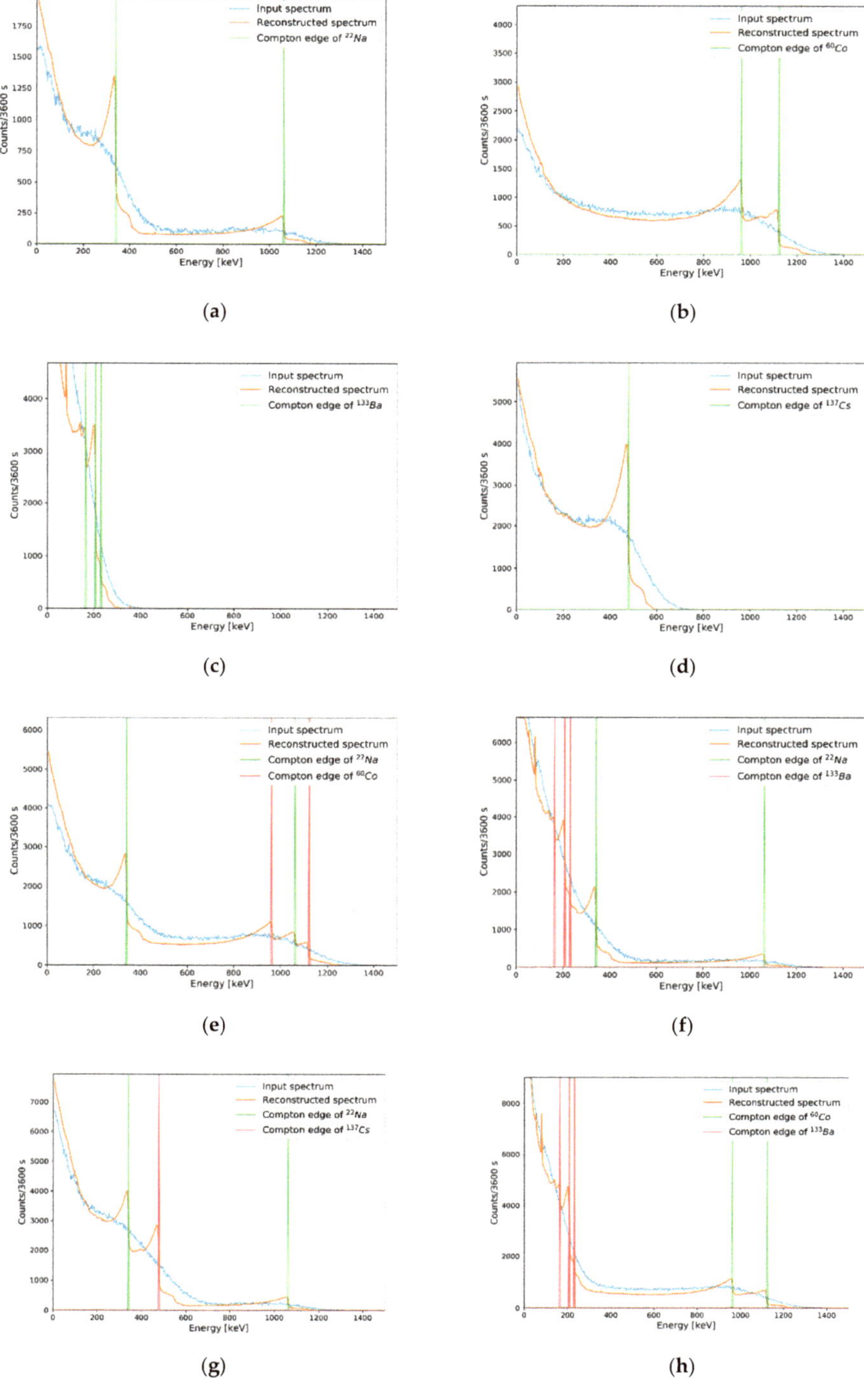

Figure 8. *Cont.*

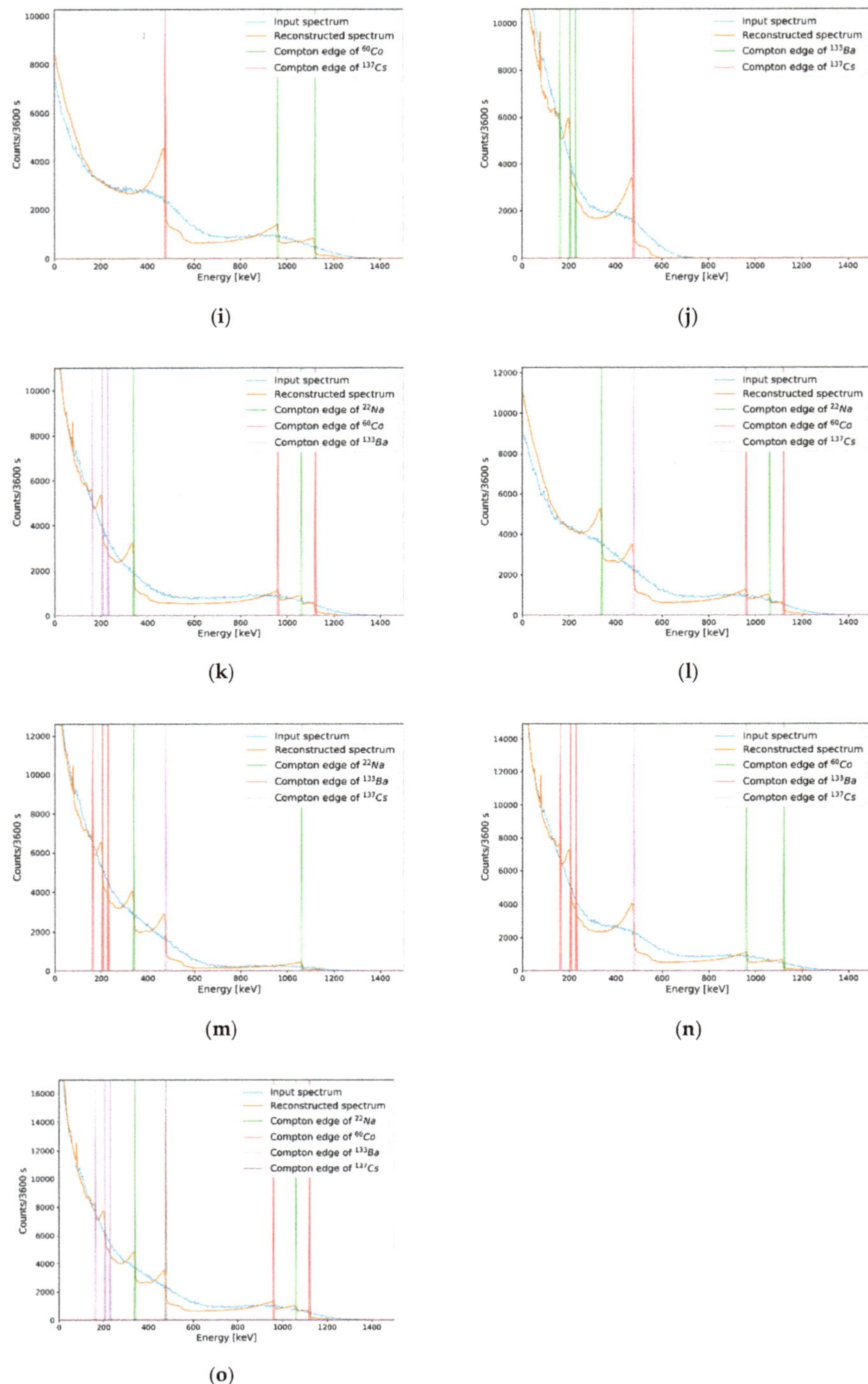

Figure 8. Results of Compton edge reconstruction for experimental data. Each figure represents

measured spectra of (**a**) ^{22}Na; (**b**) ^{60}Co; (**c**) ^{133}Ba; (**d**) ^{137}Cs; (**e**) ^{22}Na and ^{60}Co; (**f**) ^{22}Na and ^{133}Ba; (**g**) ^{22}Na and ^{137}Cs; (**h**) ^{60}Co and ^{133}Ba; (**i**) ^{60}Co and ^{137}Cs; (**j**) ^{133}Ba and ^{137}Cs; (**k**) ^{22}Na, ^{60}Co and ^{133}Ba; (**l**) ^{22}Na, ^{60}Co and ^{137}Cs; (**m**) ^{22}Na, ^{133}Ba and ^{137}Cs; (**n**) ^{60}Co, ^{133}Ba and ^{137}Cs; and (**o**) ^{22}Na, ^{60}Co, ^{133}Ba, and ^{137}Cs.

As shown in Figure 8, the energies of Compton edges in the reconstructed spectra were matched with their theoretical values calculated by Equation (4).

3.3. Minimum Reconstructible Counts

Similar to minimum detectable activity [29], the number of counts required to reconstruct Compton edges in plastic gamma spectra should be verified. In previous studies on gamma (or pseudo gamma) spectroscopy, similar concepts were defined to evaluate performance according to the activity of radioactive sources or the number of counts in their detection systems [12,30]. However, these cannot be used directly in our study because of the differences in their detailed concepts. Instead, averaged MAPE was used as a quality factor to evaluate the minimum reconstructible count (MRC) of the trained autoencoder. For each radioisotope, averaged MAPEs between reconstruction results and reconstruction references were calculated as follows. First, measured spectra in Section 3.2 (i.e., input spectra in Figure 8a–c) were normalized and utilized as PDFs for generating test sets for MRC evaluation. Second, 100 spectra were generated as test sets with the procedure detailed in Section 2.2 for each number of counts. Third, Compton edges were reconstructed for the test sets. Fourth, MAPEs between reconstruction results and reconstruction references were calculated for 100 generated spectra, and the averaged MAPE value was calculated. In this study, the reconstruction results presented in Section 3.2 (i.e., reconstructed spectra in Figure 8a–c) were used as a reconstruction reference. The threshold for MRC was determined as 10% of the averaged MAPE by referring to Table 1. Whole steps for MRC evaluation above were iterated with increment of the number of counts with interval of 50 for each radioisotope. Figure 9 shows the averaged MAPE according to the number of counts for single-isotope gamma spectra. MRCs were determined as the counts of which averaged MAPEs were decreased to less than 10%. Table 2 shows the MRCs of the single isotopes, and Figure 10 shows examples of generated spectra and reconstruction results corresponding to each MRC. In this table, MRCs are higher in order of ^{60}Co < ^{137}Cs < ^{22}Na < ^{133}Ba. The reason why MRCs are different depending on radioisotopes may be related to the intensities of energies of emitted photons and combinations of radioisotopes. ^{60}Co emits two energies of gamma rays with almost analogous ratios. However, ^{22}Na emits two energies of photons at different ratios; the intensity for a photon of 511 keV is almost double that for a photon of 1275.4 keV. This means that a higher number of counts is required to extract features for Compton edge reconstruction on the Compton continuum generated by a photon of 1275.4 keV. In the same manner, ^{133}Ba requires the highest number of counts for Compton edge reconstruction due to the complex Compton edges in the low-energy region. In the case of ^{137}Cs, the MRC was higher than the MRCs of ^{60}Co, even though it emits one energy of gamma rays. It may because higher number of counts are required to discriminate following cases; one is ^{137}Cs and the other is small ratio of ^{133}Ba and ^{137}Cs.

Table 2. Determined minimum reconstructible counts (MRCs) where averaged MAPEs are lower than 10% for each isotope.

Radioisotope	Energy [keV] [31]	Emission Intensity [%] [31]	MRC [#]
^{22}Na	511 1274.5	179.8 99.9	3050 ± 55
^{60}Co	1173.2 1332.5	99.9 99.98	650 ± 25

Table 2. *Cont.*

Radioisotope	Energy [keV] [31]	Emission Intensity [%] [31]	MRC [#]
[133]Ba	53.16	2.14	3750 ± 61
	79.61	2.65	
	80.99	32.9	
	276.4	7.16	
	302.9	18.34	
	356	62.05	
	383.8	8.94	
[137]Cs	661.66	85.21	2000 ± 44

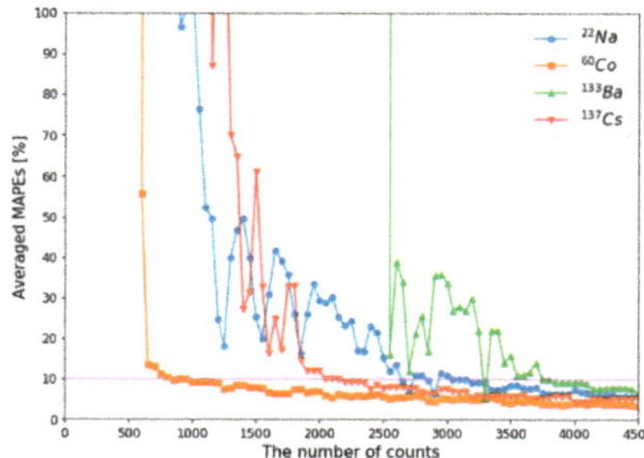

Figure 9. Averaged MAPEs according to the number of counts for single isotope gamma spectra.

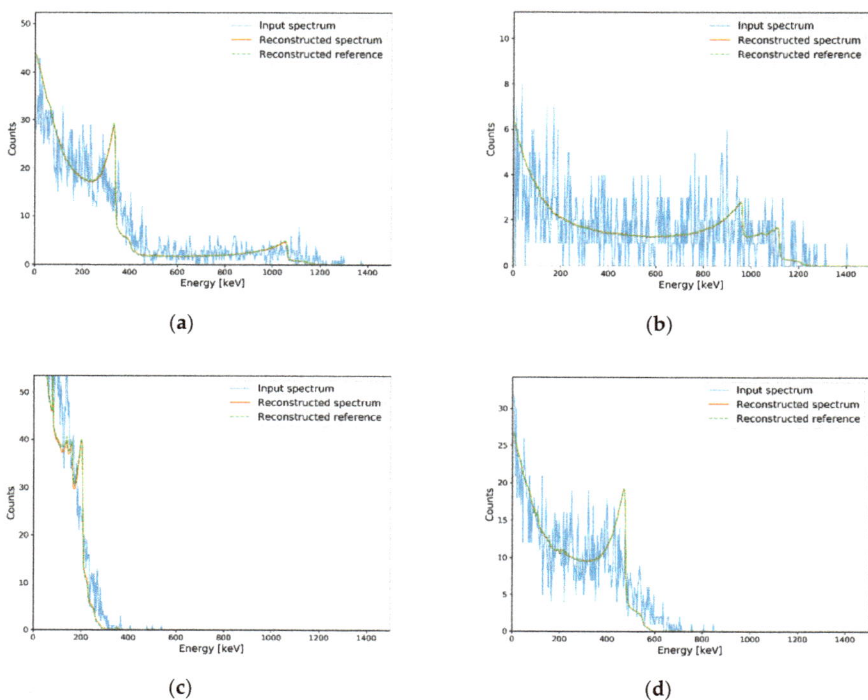

(a)

(b)

(c)

(d)

Figure 10. Examples of generated spectra for single isotope corresponding to each MRC and their

Compton edge reconstruction results. Reconstruction results on (**a**) generated spectrum for ^{22}Na, (**b**) generated spectrum for ^{60}Co, (**c**) generated spectrum for ^{133}Ba, and (**d**) generated spectrum for ^{137}Cs.

To validate the MRC evaluation results, we measured the background and each isotope for 10, 20, 40, and 80 s corresponding to MRCs of ^{60}Co, ^{137}Cs, ^{133}Ba, and ^{22}Na, respectively, and Compton edges were reconstructed from measured net spectra (i.e., background-subtracted spectra). Total net counts for each measured net spectra were not exactly the same as the MRCs but were within statistical uncertainties. Figure 11 shows the results of Compton edge reconstruction with experimental spectra for validating each MRC. Identical to Figure 8, Compton edges marked in Figure 11 were calculated by Equation (4).

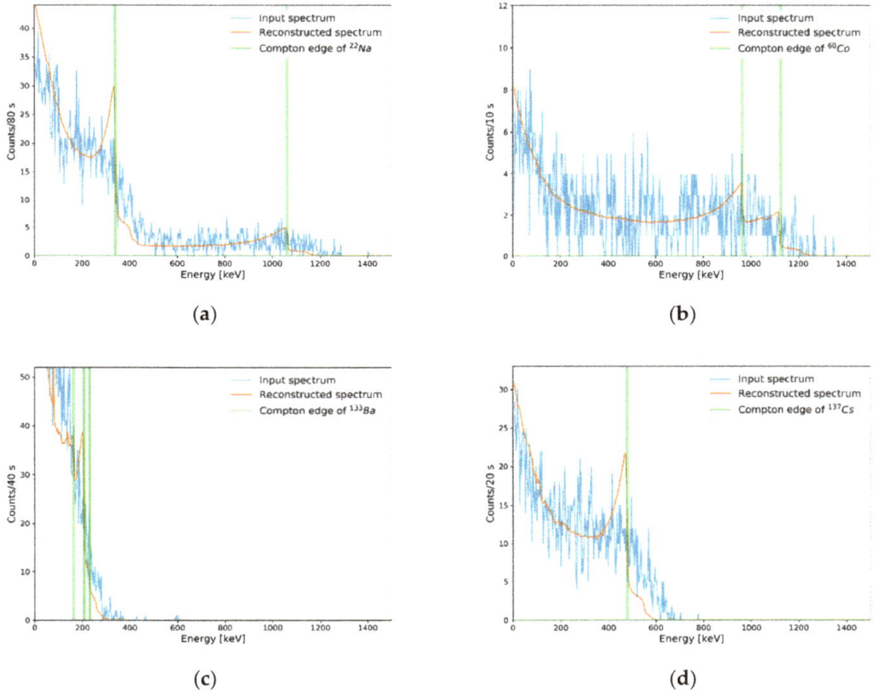

Figure 11. Results of Compton edge reconstruction for experimental spectra with different measuring periods corresponding to each MRC. Reconstruction results on (**a**) measured spectrum for ^{22}Na during 80 s, (**b**) measured spectrum for ^{60}Co during 10 s, (**c**) measured spectrum for ^{133}Ba during 40 s, and (**d**) measured spectrum for ^{133}Cs during 20 s.

4. Discussion

A deep autoencoder model was presented to reconstruct Compton edges in plastic gamma spectra. Our model was trained to reconstruct Compton edges in plastic gamma spectra, even though the spectra have poor counting statistics, by designing a dataset generation procedure. As shown by the experimental results, it successfully reconstructed Compton edges in plastic gamma spectra with statistical uncertainties. Therefore, it was possible to conduct direct pseudo gamma spectroscopy using Compton edge reconstruction results. Furthermore, the MRCs of single isotopes were evaluated with the metric of MAPE as a loss function of our model.

Although our model shows good performance on Compton edge reconstruction in plastic gamma spectra, there are three limitations we are aware of: First, the autoencoder generates data-specific results, i.e., it generates wrong results for spectra on radioisotopes that are not included in the training set; in fact, this is a characteristic of machine learning methods. For example, if untrained radioisotope is given, the autoencoder generates a spectrum which is one of the trained radioisotope or mixture of trained isotopes. Second, MRCs may be increased according to the increase in types of radioisotopes. For example, we evaluated the MRC of ^{60}Co as 650, the minimum value among three isotopes. If, however, a radioisotope emitting gamma rays of energies similar to those of ^{60}Co with almost analogous ratios was included in dataset, the MRC of ^{60}Co may be increased because more counts are required to distinguish ^{60}Co from the isotope. Furthermore, the spectra we used as input are for bare source. In practice, distortion of spectra may occur because of the presence of material surrounding the source, and it may affect Compton edge reconstruction performance. Concerning these limitations, further study is necessary.

5. Conclusions

This paper proposed a neural network model to reconstruct Compton edges in plastic gamma spectra. Datasets for training and validation of our model were generated by Monte Carlo simulations, data synthesis methods, and random sampling techniques. Although our model was trained by only simulation data, it successfully reconstructed Compton edges in simulated and measured gamma spectra, even though the spectra has poor counting statistics. Concerning the performance of Compton edge reconstruction according to counting statistics, MRCs were evaluated, and it was found that MRCs were related to the complexity of energies and intensities for emitted photons.

Many researchers have been reported methods for pseudo gamma spectroscopy such as energy windowing, F-score analysis, energy weighted, and inverse matrix algorithms. These researches excluding inverse matrix algorithm were able to find existence of radioactive materials from the patterns after spectral data processing, rather than identifying the energy of gamma rays incident on the detector. Even though inverse matrix algorithm was able to identify the energy of gamma rays from unfolded gamma-ray spectra from plastic scintillators, it does not work for spectra with poor counting statistics. However, our method allows conducting direct pseudo spectroscopy with the analysis of reconstructed Compton edges even though the spectra have poor counting statistics.

Author Contributions: Conceptualization, B.J. and Y.L.; methodology, B.J. and Y.L.; software, B.J.; validation, B.J. and Y.L.; formal analysis, B.J.; investigation, B.J.; resources, B.J. and M.M.; data curation, B.J., M.M., and J.K.; writing—original draft preparation, B.J.; writing—review and editing, G.C. and Y.L.; visualization, B.J.; supervision, G.C.; project administration, M.M.; funding acquisition, M.M. All authors have read and agreed to the published version of the manuscript.

Funding: This research was funded by the Ministry of Oceans and Fisheries, through a project titled "Development of Intelligent and Large Volume Neutron/gamma ray detection system" (KIMST project No. 20200611), and Korea Atomic Energy Research Institute through a project titled "Enhancement of operation stability of HANARO using intelligent computing technology".

Conflicts of Interest: The authors declare no conflicts of interest.

References

1. Anderson, K.K.; Jarman, K.D.; Mann, M.L.; Pfund, D.M.; Runkle, R.C. Discriminating nuclear threats from benign sources in gamma-ray spectra using a spectral comparison radio method. *J. Radioanal. Nucl. Chem.* **2008**, *276*, 713–718. [CrossRef]
2. Ely, J.H.; Kouzes, R.T.; Schweppe, J.E.; Siciliano, E.R.; Strachan, D.M.; Weier, D.R. The use of energy windowing to discriminate SNM for NORM in radiation portal monitors. *Nucl. Instrum. Meth. A* **2006**, *560*, 373–387. [CrossRef]
3. Hevener, R.; Yim, M.; Baird, K. Investigation of energy windowing algorithms for effective cargo screening with radiation portal monitors. *Radiat. Meas.* **2013**, *58*, 113–120. [CrossRef]
4. Siciliano, E.R.; Ely, J.H.; Kouzes, R.T.; Milbrath, B.D.; Schweppe, J.E.; Stromswold, D.C. Comparison of PVT and NaI(TI) scintillators for vehicle portal monitor applications. *Nucl. Instrum. Meth. A* **2005**, *550*, 647–674. [CrossRef]

5. Paff, M.G.; Fulvio, A.D.; Clarke, S.D.; Pozzi, S.A. Radionuclide identification algorithm for organic scintillatior-based radiation portal monitor. *Nucl. Instrum. Meth. A* **2017**, *849*, 41–48. [CrossRef]

6. Shin, W.; Lee, H.; Choi, C.; Park, C.; Kim, H.; Min, C. A Monte Carlo study of an energy-weighted algorithm for radionuclide analysis with a plastic scintillation detector. *Appl. Radiat. Isot.* **2015**, *101*, 53–59. [CrossRef] [PubMed]

7. Lee, H.; Shin, W.; Park, H.; Yoo, D.; Choi, C.; Park, C.; Kim, H.; Min, C. Validation of energy-weighted algorithm for radiation portal monitor using plastic scintillator. *Appl. Radiat. Isot.* **2016**, *107*, 160–164. [CrossRef] [PubMed]

8. Hamel, M.; Carrel, F. *Pseudo-gamma Spectrometry in Plastic Scintillators in New Insights on Gamma Rays*; Maghraby, A.M., Ed.; IntechOpen Limited: London, UK, 2017.

9. Ruch, M.L.; Paff, M.; Sagadevan, A.; Riviere, A.P.; Clarke, S.D.; Pozzi, S.A. Radionuclide identification by an EJ309 organic scintillator-based pedestrian radiation portal monitor using a least squares algorithm. In Proceedings of the 55th Annual Meeting of Nuclear Materials Management, Atlanta, GA, USA, 22–24 July 2014.

10. Kim, Y.; Kim, M.; Lim, K.T.; Kim, J.; Cho, G. Inverse calibration matrix algorithm for radiation detection portal monitors. *Radiat. Phys. Chem.* **2019**, *155*, 127–132. [CrossRef]

11. Kangas, L.J.; Keller, P.E.; Siciliano, E.R.; Kouzes, R.T.; Ely, J.H. The use of artificial neural networks in PVT-based radiation portal monitors. *Nucl. Instrum. Meth. A* **2008**, *587*, 398–412. [CrossRef]

12. Kim, J.; Park, K.; Cho, G. Multi-radioisotope identification algorithm using an artificial neural network for plastic gamma spectra. *Appl. Radiat. Isot.* **2019**, *147*, 83–90. [CrossRef] [PubMed]

13. Guo, G.; Pleiss, G.; Sun, Y.; Weinberger, K.Q. On calibration of modern neural networks. In Proceedings of the 34th Internal Conference of Machine Learning, International Convention Centre, Sydney, Australia, 6–11 August 2017.

14. Rumelhart, D.E.; Hinton, G.E.; Williams, R.J. Learning Internal Representations by Error Propagation. In *Parallel Distributed Processing*; MIT Press: Cambridge, MA, USA, 1986; Volume 1.

15. Oja, E. Simplified Neuron Model as a Principal Component Analyzer. *J. Math. Biol.* **1982**, *15*, 267–273. [CrossRef] [PubMed]

16. Hinton, G.E.; Salakhutdinov, R.R. Reducing the Dimensionality of Data with Neural Networks. *Science* **2006**, *313*, 504–507. [CrossRef] [PubMed]

17. Xu, H.; Chen, W.; Zhao, N.; Li, Z.; Bu, J.; Li, Z.; Liu, Y.; Zhao, Y.; Pei, D.; Feng, Y.; et al. Unsupervised Anomaly detection via Variational Auto-Encoder for Seasonal KPIs in Web Applications. In Proceedings of the 2018 World Wide Web Conference, Lyon, France, 23–27 April 2018; pp. 187–196.

18. Sakurada, M.; Yairi, T. Anomaly Detection Using Autoencoders with Nonlinear Dimensionality Reduction. In Proceedings of the MLSDA 2014 2nd Workshop on Machine Learning for Sensory Data Analysis, Gold Coast, Australia, 2 December 2014; pp. 4–11.

19. Vincent, P.; Larochelle, H.; Lajoie, I.; Bengio, Y.; Manzagol, P.-A. Stacked Denoising Autoencoders: Learning Useful Representations in a Deep Network with a Local Denoising Criterion. *J. Mach. Learn. Res.* **2010**, *11*, 3371–3408.

20. Datasheet of R2228, HAMAMATSU Photonics K.K. Available online: https://www.hamamatsu.com/resources/pdf/etd/R2228_TPMH1062E.pdf (accessed on 7 May 2020).

21. Datasheet of E990-501, HAMAMATSU Photonics K.K. Available online: https://www.hamamatsu.com/resources/pdf/etd/PMT_90-101_en.pdf (accessed on 7 May 2020).

22. Jeon, B.; Kim, J.; Moon, M.; Cho, G. Parametric optimization for energy calibration and gamma response function of plastic scintillation detectors using a genetic algorithm. *Nucl. Instrum. Meth. A* **2019**, *930*, 8–14. [CrossRef]

23. Werner, C.J.; Bull, J.S.; Solomon, C.J.; Brown, F.B.; McKinney, G.W.; Rising, M.E.; Dixon, D.A.; Martz, R.E.; Hughes, H.G.; Cox, L.J.; et al. *MCNP6.2 Release Notes*; Report LA-UR-18-20808; Loa Alamos National Laboratory: Los Alamos, NM, USA, 2018.

24. McConn, R.J., Jr.; Gesh, C.J.; Pagh, R.T.; Rucker, R.A.; Williams, R.G., III. *Compendium of Material Composition Data for Radiation Transport Modeling*; Technical Report PNNL-15870, Rev 1; Pacific Northwest National Laboratory: Richard, WA, USA, 2011.

25. Abadi, M.; Agarwal, A.; Barham, P.; Bervdo, E.; Chen, Z.; Citro, C.; Corrado, G.S.; Davis, A.; Dean, J.; Devin, M.; et al. TenforFlow: Large-Scale Machine Learning on Heterogeneous Systems. 2015. Available online: https://www.tensorflow.org (accessed on 20 April 2020).

26. Chollet, F. KERAS. 2015. Available online: https://keras.io. (accessed on 20 April 2020).

27. Zeiler, M.D. ADADELTA: An Adaptive Learning Rate Method. Method. *arXiv* **2012**, arXiv:1212.5701.

28. Knoll, G.F. Radiation Interactions. In *Radiation Detection and Measurement*, 3rd ed.; Zobrist, B., Hepburn, K., Factor, R., Malinowski, S., Lesure, M., Eds.; John Wiley & Sons, Inc.: Hoboken, NJ, USA, 2000; p. 51.

29. Kirkpatric, J.M.; Venkataraman, R.; Young, B.M. Minimum detectable activity, systematic uncertainties, and the ISO 11929 standard. *J. Radioanal. Nucl. Chem.* **2013**, *296*, 1005–1010. [CrossRef]

30. Keyser, R.M.; Sergent, F.; Twomey, T.R.; Upp, D.L. Minimum detectable activity estimates for a germanium-detector based spectroscopic portal monitor. In Proceedings of the Institute of Nuclear Materials Management 47th Annual Meeting, Nashville, TN, USA, 16–20 July 2006.

31. Decay Radiation Search from National Nuclear Data Center in Brookhaven National Laboratory. Available online: https://www.nndc.bnl.gov/nudat2/index_dec.jsp (accessed on 7 May 2020).

Article

Uncertainty Estimation of the Dose Rate in Real-Time Applications Using Gaussian Process Regression

Jinhwan Kim [1], Kyung Taek Lim [1], Kyeongjin Park [1], Yewon Kim [2] and Gyuseong Cho [1,*]

[1] Department of Nuclear and Quantum Engineering, Korea Advanced Institute of Science and Technology, 291, Daehak-ro, Yuseong-gu, Daejeon 34141, Korea; kjhwan0205@kaist.ac.kr (J.K.); kl2548@kaist.ac.kr (K.T.L.); myesens@kaist.ac.kr (K.P.)

[2] The Center for Nuclear Nonproliferation Strategy and Technology, Korea Institute of Nuclear Nonproliferation and Control, 1534 Yuseong-daero, Yuseong-gu, Daejeon 34054, Korea; yewonkim@kinac.re.kr

*   Correspondence: gscho@kaist.ac.kr

Received: 16 April 2020; Accepted: 13 May 2020; Published: 19 May 2020

Abstract: Major standard organizations have addressed the issue of reporting uncertainties in dose rate estimations. There are, however, challenges in estimating uncertainties when the radiation environment is considered, especially in real-time dosimetry. This study reports on the implementation of Gaussian process regression based on a spectrum-to-dose conversion operator (i.e., G(E) function), the aim of which is to deal with uncertainty in dose rate estimation based on various irradiation geometries. Results show that the proposed approach provides the dose rate estimation as a probability distribution in a single measurement, thereby increasing its real-time applications. In particular, under various irradiation geometries, the mean values of the dose rate were closer to the true values than the point estimates calculated by a G(E) function obtained from the anterior–posterior irradiation geometry that is intended to provide conservative estimates. In most cases, the 95% confidence intervals of uncertainties included those conservative estimates and the true values over the range of 50–3000 keV. The proposed method, therefore, not only conforms to the concept of operational quantities (i.e., conservative estimates) but also provides more reliable results.

Keywords: spectrum-to-dose conversion operator; G(E) function; gaussian process regression; dose rate uncertainty; real-time dosimetry; operational quantities

---

1. Introduction

The concepts of equivalent dose and effective dose were first introduced by the International Commission on Radiological Protection (ICRP) in order to provide recommendations and guidelines for the protection of people and the environment in an integrated manner in all exposure situations [1–3]. However, given that these concepts are not measurable quantities, the International Commission on Radiological Units and Measurements (ICRU) defined a few measurable operational quantities to establish convenient and appropriate evaluations of an equivalent and effective dose [4,5]. In cases of strongly penetrating radiations, such as gamma rays and neutrons, an adequate operational quantity for monitoring a specific area is defined by the ambient dose equivalent H*(10) (hereafter referred to as "ambient dose rate" and used interchangeably with the term "dose rate"). The ICRU defined the ambient dose rate as, "The dose equivalent at a point in a radiation field that would be produced by a corresponding expanded and aligned field in the ICRU sphere at a depth of 10 mm on the radius vector opposing the direction of the aligned field."

The response of the ambient dose rate is highly dependent on photon energy and the angle of radiation incidence. Therefore, the requirement of IEC 60846:2009 advises that the relative response of the dose rate to the reference radiation (e.g., Cs-137) within the combined rate range of photon energy and the angle of incidence shall be between 0.6 and 1.4 [6]. Intending to minimize the effect of photon energy on dose rate response and to achieve more accurate estimates of the dose rate, the use of scintillation detectors is a way to make the response less sensitive to radiation energy by obtaining an energy spectrum. The $G(E)$ function is a typical example of conversion from the energy spectrum to dose rate [7–12]; the measured spectrum is directly converted into dose rate without applying stripping or unfolding methods [13–15]. This method is, therefore, often adopted for real-time dose measurement. For the estimation of the $G(E)$ function, the most conservative direction of irradiation, anterior posterior, is typically assumed rather than those in other idealized geometries, such as rotational and isotropic ones. However, in a real contaminated environment, the irradiation direction of photons entirely depends on typically unknown source distributions. Although isotropic or rotational geometry approximate certain real irradiation conditions [9,16], these are not the same as the idealized ones [17]. In addition, since the response of dosimetry is normalized to the ambient dose rate under one of the geometric conditions, most errors in dose rate estimation primarily arise from the calculation of the dose conversion operator. From a safety standpoint, it is necessary to provide conservative dose rate estimates. In this respect, an alternative is the use of the maximum value for the dose conversion operator, which would be similarly produced by various irradiation geometries, as already proposed [18]. However, it is often more important to report the best estimate and the best evaluation of dose rate uncertainty that includes a conservative estimate. Therefore, a different approach would be preferable to ensure that the dose rate is presented with the best estimate of the mean and its expending uncertainty (i.e., 1.96 standard deviations) for real-time applications.

This study presents a new spectrum-to-dose conversion operator concept, called $G(E)_{GPR}$ functions, which are $G(E)$ functions based on Gaussian process (GP) regression that account for the relative response to radiation energy and direction of radiation incidence in order to deal with uncertainty in dose rate estimation. A GP model can be constructed using all the data points of $G(E)$ functions determined under various irradiation geometries, e.g., the angles of incidence of 0°, 45°, and 90°, and isotropic geometry. Then, a set of $G(E)_{GPR}$ functions can be obtained through independent realizations (or equivalently, sample path) of the GP model, where each realization defines a conversion factor for every possible energy step. Lastly, the obtained $G(E)_{GPR}$ functions are multiplied by an observed spectrum to make it possible to estimate the mean dose rate value and its associated uncertainty. Figure 1 illustrates the proposed concept in comparison with the conventional method. This paper presents simulation results that demonstrate the behavior and performance of the proposed approach. It is worth noting that although this study focuses on the ambient dose rate, this method can be applied to any dosimetric quantity (e.g., air kerma) if the target quantity is defined as a function of radiation energy.

2. Materials and Methods

2.1. G(E) Function

A general description of the $G(E)$ function in terms of the dose conversion coefficient $h(E_0)$ at mono-energy E_0 and the response function of a detector $R(E, E_0)$, which represents the photon of energy E_0 depositing energy E into the detector, can be expressed as

$$h(E_0) = \int_{E_{min}}^{E_{max}} R(E, E_0)G(E)dE, \tag{1}$$

where E_{min} and E_{max} are the minimum and maximum detectable energies deposited in the detector, respectively. The total dose rate (D) in multi-energy radiation conditions can, therefore, be represented as

$$
\begin{aligned}
D &= \sum_i \varnothing(E_i)h(E_i) \\
&= \sum_i \varnothing(E_i) \int_{E_{min}}^{E_{max}} R(E, E_i)G(E)dE \\
&= \int_{E_{min}}^{E_{max}} \sum_i \varnothing(E_i)R(E, E_i)G(E)dE \\
&= \int_{E_{min}}^{E_{max}} M(E)G(E)dE.
\end{aligned}
\tag{2}
$$

where $\varnothing(E_i)$ is the fluence rate at the energy of E_i. In real conditions, the integral of continuous energies should be changed to the sum of discrete energies (or equivalently, the number of channels N; in this case, $N = 869$).

$$
D = \sum_{i=1}^{N} M(E_i)G(E_i)
\tag{3}
$$

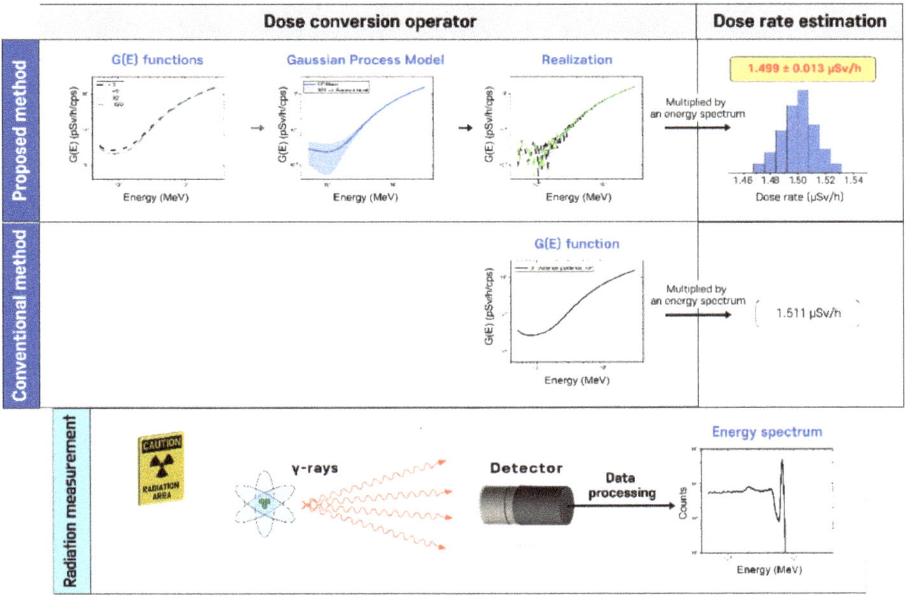

Figure 1. Illustration of the proposed concept of dealing with uncertainty in dose rate estimation compared with the conventional method.

Consequently, the total dose rate can be directly estimated using the $G(E)$ function and a measured spectrum. According to related studies [7–9,11], the $G(E)$ function can be expressed as

$$
G(E) = \sum_{K=1}^{K_{max}} A(K)(log(E))^{K-M-1},
\tag{4}
$$

where $A(K)$ is a parameter, K_{max} is the number of terms, and M is constant? The values for K_{max} and M were set to 7 and 0, respectively. It should be noted that the optimization of these parameters is not the main concern of this study. Dose rate can be represented by combining Equations (3) and (4):

$$D = \sum_{i=1}^{N} M(E_i) \sum_{K=1}^{K_{max}} A(K)(\log(E_i))^{K-M-1}. \tag{5}$$

To compute $A(K)$, it is required to obtain spectra with known mono or multiple energies and the corresponding dose rates. The availability of energy sources limits actual experiments; however, Monte Carlo simulations allow for the use of any energy, so corresponding dose rates can be calculated, given that detector geometry has been properly defined. Finally, $A(K)$ were obtained using the gradient descent method [11].

2.2. GP Regression

This section briefly introduces the concept of GP regression employed for implementation purposes to deal with uncertainty in dose rate estimation. More details can be found in [19,20].

The GP is expressed as a distribution over functions for which any finite subset of variables has a joint multivariate Gaussian distribution. Since the GP is described by Gaussian distribution, it is parameterized by its mean function $m(x)$ and positive definite covariance function $k(x, x')$, also known as a kernel function:

$$f(\mathbf{x}) \sim \mathcal{GP}(m(\mathbf{x}), k(\mathbf{x}, \mathbf{x}')) \tag{6}$$

Typically, $m(x)$ is set to 0 to avoid expensive computations in posterior distribution and make inferences only via the kernel function. The kernel function takes two indices x and x' and returns their corresponding modeled covariance. By choosing an appropriate kernel function, it is possible to incorporate assumptions such as smoothness and likely patterns that are expected in the data. A popular choice of the kernel is the radial basis function kernel, where two points are exponentially correlated, depending on the distance between them.

The main assumption in GP modeling is that output y is an observation of $f(x)$ that has been corrupted by Gaussian noise ϵ:

$$y = f(x) + \varepsilon, \ \varepsilon \ \sim \ \mathcal{N}\!\left(0, \sigma_\varepsilon^2\right) \tag{7}$$

where noise term ϵ is assumed independent and identically distributed with zero means. Hence $f(x)$ is a latent variable whose posterior distribution will be inferred after observing new samples at various locations in the domain. The resultant inference is called GP regression.

Suppose training outputs y_t have been observed and predictions for test outputs f_* have been made. They then follow a joint normal distribution:

$$\begin{bmatrix} y_t \\ f_* \end{bmatrix} \sim \mathcal{N}\!\left(0, \begin{bmatrix} K(X_t, X_t) + \sigma_\varepsilon^2 I & K(X_t, X_*) \\ K(X_*, X_t) & K(X_*, X_*) \end{bmatrix}\right) \tag{8}$$

where X_t and X_* are the design matrices for training and test data, respectively. $K(X_t, X_t)$ represents the covariance matrix between all points observed so far in the training data, which is similarly true for other covariance matrices of $K(X_t, X_*)$, $K(X_*, X_t)$, and $K(X_*, X_*)$. I is an identity matrix whose diagonal elements and off-diagonal elements are 1 and 0, respectively? Conditioning f_* on the observation y_t $p(f_* | X_t, y_t, X_*)$, the predictive distribution of test points with respect to the mean and covariance matrix can be written as:

$$m_t(x) = K(x, X_t)\left[K(X_t, X_t) + \sigma_\varepsilon^2 I\right]^{-1} y_t \tag{9}$$

$$K_t(x, x') = K(x, x') - K(x, X_t)\left[K(X_t, X_t) + \sigma_\varepsilon^2 I\right]^{-1} K(X_t, x') \tag{10}$$

Consequently, the estimation of the posterior mean and covariance are involved in calculating four different covariance matrices.

2.3. Monte Carlo Modeling and Simulation

A Monte Carlo N-Particle Transport Code (MCNP6) [21] was used to validate the proposed method. A schematic of the MCNP6 model used for simulations is illustrated in Figure 2. The 5.08 × 5.08 cm (diameter × height) NaI(Tl) crystal with a density of 3.6 g cm^{-3} was covered by a MgO reflector with a density of 2 g cm^{-3}, which was surrounded by aluminum with a density of 2.7 g cm^{-3}. A 20-mm-thick aluminum plate was placed behind the crystal to mimic a phenomenon where photons are scattered or backscattered in a photomultiplier tube [22]. This is a reasonable assumption because it considers the scattered effects of the photomultiplier tube on a spectrum as the angle of irradiation direction changes. A parallel beam of photons distributed over a circular source was irradiated on the NaI(Tl) detector. In order to obtain $G(E)_g$ functions under various directions of irradiation, the directions of the detector were rotated against the circular source by specific angle α (i.e., 0, 45, and 90), where subscript g is the irradiation geometry that determines the $G(E)$ function. For isotropic geometry, the detector was centered inside the spherical source, emitting fully isotropic irradiation of photons.

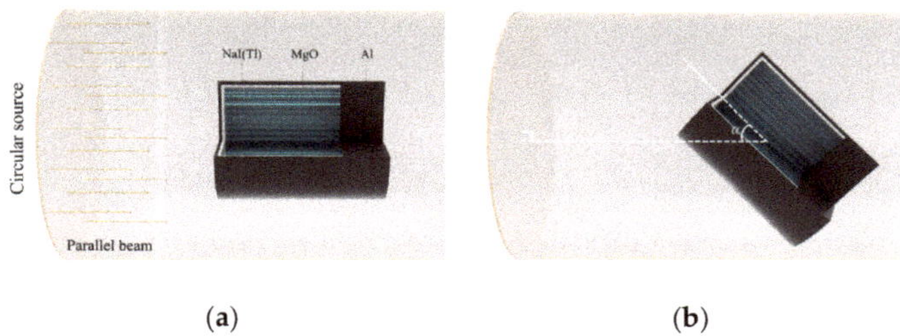

Figure 2. Schematic of the calculation geometry defined for MCNP6 simulations. (**a**) A parallel photon beam was irradiated on an NaI(Tl) detector. (**b**) The directions of the detector were rotated against the circular source by specific angle α. A 20 mm-thick aluminum disk was positioned on the back of the crystal to consider the scattering in a photomultiplier tube.

Since an actual spectrum is influenced by the broadening effect due to the statistical variation of the scintillation light signals and various electronic sources of noise, a simulated spectrum must be modified to accommodate such effects. This can be regarded as a convolution process of an ideal spectrum with the kernel of a broadening filter. A Gaussian-energy broadening filter is often applied, meaning that a delta function type of peak becomes a Gaussian function with full-width at half-maximum value (FWHM = 2.36 × sigma). In the MCNP, a non-linear function with three parameters regarding FWHM is specified to apply broadening effects on the ideal spectrum. The optimal values of parameters obtainable from measured spectra were found using a genetic algorithm [23].

3. Results

3.1. G(E) Functions for Idealized Irradiation Geometires

Figure 3 shows the determined $G(E)$ functions of the NaI (Tl) detector as a function of energy deposited in the crystal under four different irradiation geometries. As expected, the $G(E)_0$ function, i.e., $G(E)$ function for the angle of incidence of 0, tended to yield higher values over the entire energy range than those for other directions of photons with respect to the detector axis. Additionally, there were relatively small differences between the values of $G(E)$ functions for energies above 200 keV, showing good agreement with previous results [9,12]. This could be ascribed to the fact that the energy deposition of relatively high energy depends primarily on the volume of the crystal. On the other hand, the values of the $G(E)$ functions obtained from different types of irradiation geometry tended to relatively disperse, especially for energies below 200 keV. This is because the photons in that energy range have high interaction probabilities with the crystal, so the energy deposition in the crystal becomes proportional to the projected area of the crystal incident surface. These results suggest that the estimated dose rate may drift from the true value, especially in the low energy range, depending on the $G(E)$ functions calculated by different types of irradiation geometry.

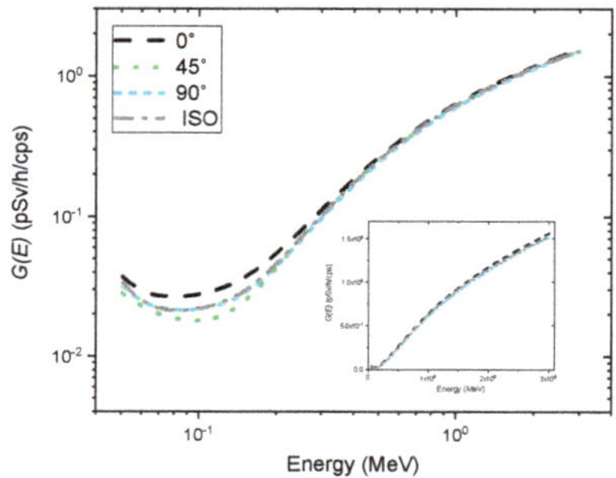

Figure 3. Spectrum-to-dose conversion operator (i.e., $G(E)$ functions) for the angles of incidence of 0 (black dashed line), 45 (green dotted line), and 90 (cyan short-dashed line), and isotropic geometry (gray dash-dotted line). The inset shows the same graph on a linear scale.

3.2. G(E) Functions Using GP Regression

Figure 4 shows the posterior mean of the GP model as well as its probabilistic nature in the form of a 95% confidence interval using the data from previously-determined $G(E)$ functions; the $G(E)$ functions are also illustrated in this figure for comparison. The result shows that the GP model no longer has a single value for energy but a distribution (i.e., Gaussian distribution) indexed by energy. In addition, the entire data points of $G(E)$ functions determined under different types of irradiation geometry were found inside the 95% confidence region of the posterior. In particular, the relative vertical width of the confidence region with regard to the mean value tended to increase as it moved to the low energy range, especially for energies below 200 keV, to account for variations induced by radiation energy and direction of radiation incidence. It should be noted that the absolute values of the confidence region obtained from the GP model over the entire energy range are almost similar.

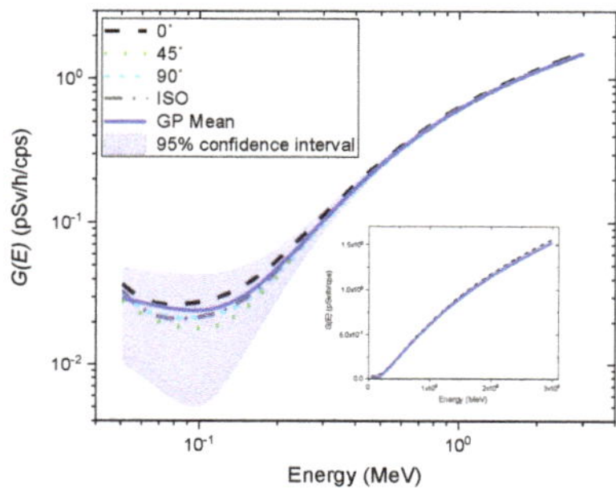

Figure 4. Gaussian process (GP) regression using the data from the previously-determined $G(E)$ functions under the angles of incidence of 0° (black dashed line), 45° (green dotted line), and 90° (cyan short-dashed line), and isotropic geometry (gray dash-dotted line). The blue solid line represents the mean of the GP model. The blue shaded area denotes a 95% confidence interval. The previously determined $G(E)$ functions are also illustrated for comparison. The inset shows the same graph on a linear scale.

Figure 5 shows an example of independent realization functions (i.e., $G(E)_{GPR}$ functions) randomly sampled from the GP model. As expected, each $G(E)_{GPR}$ function represented a different path because of the randomness of the stochastic process, fluctuating around the mean of the GP model. This implies that multiple dose rate values can be calculated using the $G(E)_{GPR}$ functions multiplied by the observed spectrum, resulting in not only the best dose rate estimate but also its uncertainty, which might contain the true value. The mean of the GP model was lower than that of the $G(E)_0$ function, which generally overestimates dose rates, so it nearly coincided with the $G(E)_{ISO}$ function (see Figure 4). That is, the mean dose rate values and the dose rates estimated by the $G(E)_{ISO}$ function might be in good agreement. This result is quite promising because isotropic geometry can be a reasonable assumption for irradiations often received from naturally occurring radioisotopes in homes or the surrounding environments [9,16].

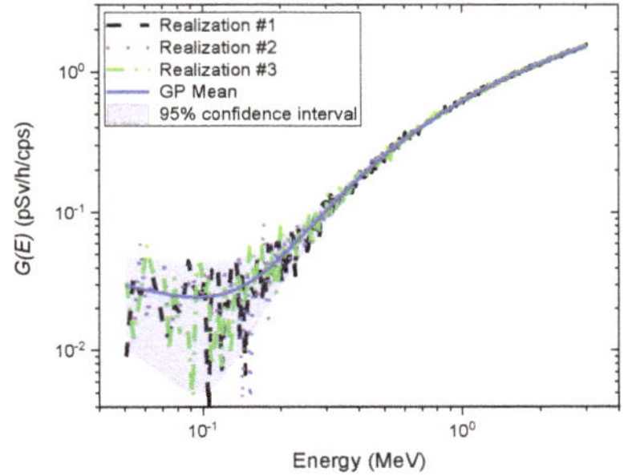

Figure 5. Example of three independent realization functions (i.e., $G(E)_{GPR}$ functions) randomly sampled from the GP model. The solid blue line represents the mean of the GP model. The blue shaded area denotes the 95% confidence interval.

3.3. Dose Rate Uncertainty Estimation

To validate the proposed method, various spectra were obtained for mono-energy over the range of 50–3000 keV at certain intervals with various geometries, e.g., the angles of incidence of 0°, 45°, and 90°, and isotropic geometry. To calculate the uncertainty of the dose rate (i.e., 95% confidence interval), 100 $G(E)_{GPR}$ functions were randomly sampled from the GP model each time. Figure 6 shows a comparison of the energy response normalized to the energy of 622 keV emitted by Cs-137, estimated with $G(E)_{GPR}$ functions and the $G(E)_0$ function for the spectra obtained at the angle of incidence of 0°. Here, the energy response was calculated by having the ratio of the estimated dose rate to the true dose rate at certain energy divided by the same ratio at the energy of 662 keV. That is, an increase in the value of the energy response suggests that the dose rate is overestimated, or vice versa. As we can see from the figure, the energy responses for the $G(E)_0$ function were reasonably close to one, which means that the estimated values of the dose rate and true values were in good agreement. This is because the test spectra were acquired under the same condition used for the $G(E)_0$ function calculation. These results are not as good as those that were reported by previous studies, especially for the low energy range. Nonetheless, it is worth emphasizing that the purpose of this study was to propose a concept that would make it possible to deal with uncertainty existing in the dose rate by taking into account the relative response of radiation energy and the direction of radiation incidence. For the dose rates estimated with $G(E)_{GPR}$ functions, the mean values deviated slightly more from the reference value 0 for energies below 200 keV. The reason is that there might be a discrepancy between the mean values of $G(E)_{GPR}$ functions sampled from the GP model and those from the $G(E)_0$ function, which is better suited with respect to the test spectra. In contrast, the proposed method was able to provide the uncertainty and the mean of the dose rate. As expected, the relative uncertainty tended to increase with a decrease in energy. In particular, the relative uncertainty increased sharply for energies under 200 keV because the GP model had relatively wide intervals for that energy range. In addition, the 95% confidence interval of relative uncertainty mostly included the true value and the conservative values obtained with the G(E)$_0$ function.

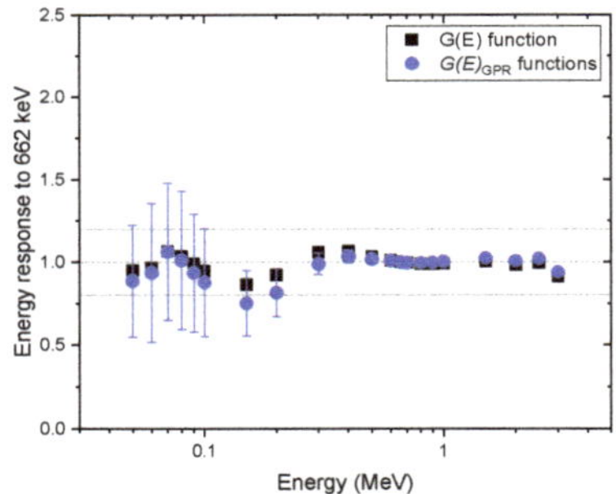

Figure 6. Comparison of energy response normalized to the energy of 622 keV, estimated with $G(E)_{GPR}$ functions, and the G(E)$_0$ function for the spectra obtained under the angle of incidence of 0. The error bar shows a 95% confidence interval.

Figures 7–9 show the same comparison of energy response for the spectra obtained at the angles of incidence of 45 and 90, and isotropic geometry. Although similar trends were observed in the non-zero angle of incidences, the estimated dose rate values obtained with the G(E)$_0$ function were overestimated, especially for energies below 600 keV for all geometries in which the test spectra were acquired, which was expected. This shows the reason that the G(E)$_0$ function is generally used to provide conservative dose estimates. In particular, the dose rate overestimation for the test spectra, assuming that photons were irradiated under the angle of incidence of 45°, is as high as 50% at 70 keV (see Figure 7). This is because the largest projected area of the crystal incident surface is generated at that specific angle, which increases interaction probabilities, especially for photons at low energies. Likewise, the mean values of the dose rate estimate, with $G(E)_{GPR}$ functions, showed similar trends but were less overestimated. Furthermore, the uncertainties of those estimates included not only the conservative values estimated by the G(E)$_0$ function but also the true values in most situations. The uncertainty calculated by the proposed method is much more reliable than those that provide only a point estimate, because in the real world, it is not possible to know how far an estimate is from the true value.

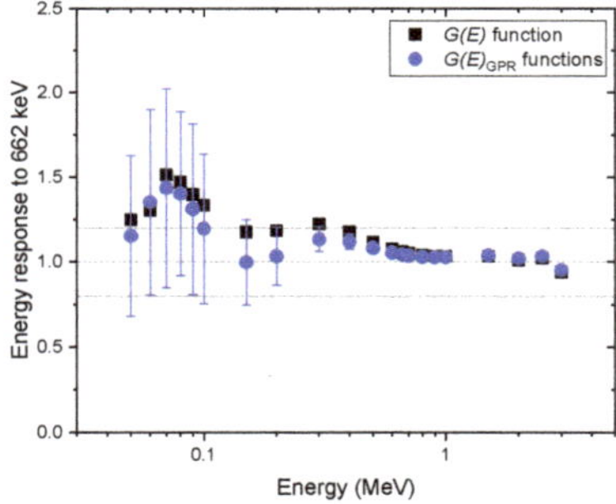

Figure 7. Comparison of energy response normalized to the energy of 622 keV, estimated with $G(E)_{GPR}$ functions, and the G(E)$_0$ function for the spectra obtained under the angle of incidence of 45. The error bar shows a 95% confidence interval.

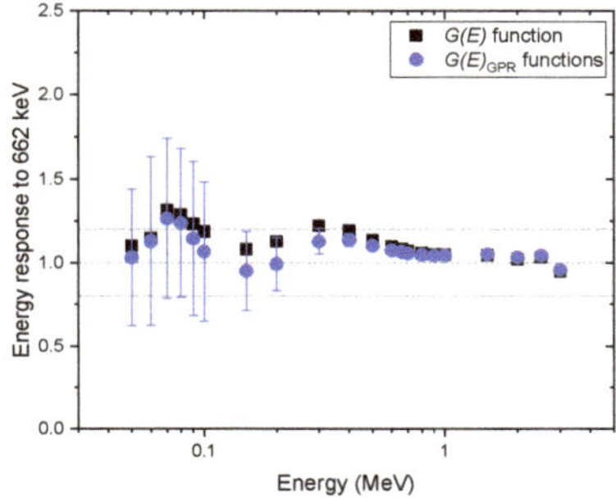

Figure 8. Comparison of energy response normalized to the energy of 622 keV, estimated with $G(E)_{GPR}$ functions, and the $G(E)_0$ function for the spectra obtained under the angle of incidence of 90. The error bar shows a 95% confidence interval.

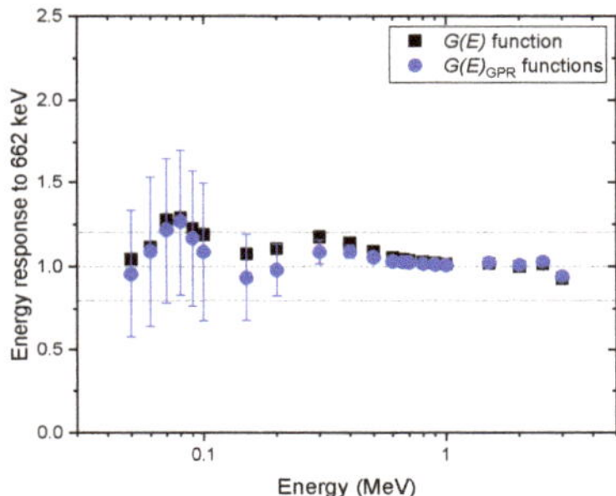

Figure 9. Comparison of energy response normalized to the energy of 622 keV, estimated with $G(E)_{GPR}$ functions, and the $G(E)_0$ function for the spectra obtained under isotropic geometry. The error bar shows a 95% confidence interval.

4. Discussion

This paper presented how GP regression can be applied to deal with dose rate uncertainty in real-time applications. The results demonstrated that the proposed approach is much more reliable and robust in comparison with existing methods. For conventional methods, a way to determine uncertainties is the statistical analysis of a series of observations (i.e., Type A uncertainties). In this case, however, they ignore other components of uncertainty determined by scientific judgment based on published data (e.g., $G(E)$ function). Furthermore, in cases of real-time applications, the estimation of Type A uncertainties is practically impossible, so the uncertainty associated with dose rate can, therefore, not be reported. Although previous studies attempted to estimate dose rate uncertainty, they simply averaged $G(E)$ functions for the angles of incidence of $0°$ and $90°$ over the entire energy range and neglected the combined effects of the direction of radiation incidence and photon energy on the detector response [24]. In contrast, this study constructed a GP model that considers the relative responses to various irradiation geometries and energy to provide the mean and its uncertainty for the estimated dose rate based on a single spectrum. In addition, the mean dose rate values were not heavily overestimated under various irradiation geometries, and the 95% confidence interval of uncertainties included the conservative estimates obtained with the $G(E)_0$ function and true values. An estimated value without a statement about its associated uncertainty is less informative because it does not make it possible to quantify the potential risk arising from radiation exposure and does not indicate the precision of the estimate. Lastly, the calculated uncertainty allows for a quantitative comparison with results reported by other investigators, which enables its assessment.

5. Conclusions

This work presented a new method for dose rate estimation that uses GP regression. The presented results confirmed that numerous $G(E)_{GPR}$ functions that account for the relative responses to radiation energy and irradiation directions could be randomly sampled from a GP model, making it possible to deal with uncertainty in dose rate estimation for real-time applications. While the conventional method overestimates the dose rate by as much as 50% under different irradiation geometries, the mean values of the dose rate estimated with $G(E)_{GPR}$ functions were closer to the true value. Furthermore, the overestimated values obtained with the $G(E)_0$ function and the true values were mostly found within the 95% confidence interval of uncertainty. Therefore, the proposed method conforms to the concept of operational quantities present in conservative estimates and provides a more reliable dose rate estimation.

Author Contributions: Conceptualization, J.K. and G.C.; methodology, J.K. and K.T.L.; software, J.K.; validation, J.K. and K.T.L.; formal analysis, J.K. and K.P.; investigation, J.K.; writing—original draft preparation, J.K.; writing—review and editing, G.C. and K.T.L.; visualization, J.K. and Y.K.; supervision, G.C.; funding acquisition, G.C. All authors have read and agreed to the published version of the manuscript.

Funding: This research was supported by the Korea Atomic Energy Research Institute Funded by the Ministry of Science and ICT (2020M2C9A1068162) and NRF (National Research Foundation of Korea) Grant funded by the Korean Government (NRF-2018-Global Ph.D. Fellowship Program).

Conflicts of Interest: The authors declare no conflict of interest.

References

1. ICRP. *International Commission on the ICRP*; ICRP Publication 26; ICRP Pergamon: Oxford, UK, 1977; Available online: https://journals.sagepub.com/doi/pdf/10.1177/ANIB_1_3 (accessed on 19 May 2020).
2. ICRP. *1990 Recommendations of the International Commission on Radiological Protection*; ICRP Publication 60; ICRP Pergamon: Oxford, UK, 1991; Available online: https://journals.sagepub.com/doi/pdf/10.1177/ANIB_21_1-3 (accessed on 19 May 2020).
3. ICRP. *The 2007 Recommendations of the International Commission on Radiological Protection*; ICRP Publication 103; 2007; Available online: https://journals.sagepub.com/doi/pdf/10.1177/ANIB_37_2-4 (accessed on 19 May 2020).
4. Zaider, M. ICRU, Determination of Dose Equivalents Resulting from External Radiation Sources. *Radiat. Res.* **1989**, *120*, 375. [CrossRef]
5. ICRU. *Determination of Dose Equivalents from External Radiation Sources—Part II, Report 43*; 1988; Available online: https://icru.org/home/reports/determination-of-dose-equivalents-from-external-radiation-sources-part-ii-report-43 (accessed on 19 May 2020).
6. IEC. *Radiation Protection Instrumentation—Ambient and/or Directional Dose Equivalent (Rate) Meters and/or Monitors for Beta, X and Gamma Radiation—Part 1: Portable Workplace and Environmental Meters and Monitors*; 2009; Available online: https://webstore.iec.ch/publication/3682 (accessed on 19 May 2020).
7. Terada, H.; Sakai, E.; Katagiri, M. Spectrum-to-Exposure Rate Conversion Function of a Ge(Li) in-Situ Environmental Gamma-Ray Spectrometer. *IEEE Trans. Nucl. Sci.* **1977**, *24*, 647–651. [CrossRef]
8. Huang, P. Measurement of air kerma rate and ambient dose equivalent rate using the G(E) function with hemispherical CdZnTe detector. *Nucl. Sci. Tech.* **2018**, *29*, 35. [CrossRef]
9. Tsuda, S.; Saito, K. Spectrum–dose conversion operator of NaI(Tl) and CsI(Tl) scintillation detectors for air dose rate measurement in contaminated environments. *J. Environ. Radioact.* **2017**, *166*, 419–426. [CrossRef]
10. Tsutsumi, M.; Tanimura, Y. LaCl3(Ce) scintillation detector applications for environmental gamma-ray measurements of low to high dose rates. *Nucl. Instrum. Methods Phys. Res. Sect. A Accel. Spectrometers Detect. Assoc. Equip.* **2006**, *557*, 554–560. [CrossRef]
11. Park, K.; Kim, J.; Lim, K.T.; Kim, J.; Chang, H.; Kim, H.; Sharma, M.; Cho, G. Ambient dose equivalent measurement with a CsI(Tl) based electronic personal dosimeter. *Nucl. Eng. Technol.* **2019**, *51*, 1991–1997. [CrossRef]
12. Ji, Y.-Y.; Chung, K.H.; Lee, W.; Park, D.-W.; Kang, M.J. Feasibility on the spectrometric determination of the individual dose rate for detected gamma nuclides using the dose rate spectroscopy. *Radiat. Phys. Chem.* **2014**, *97*, 172–177. [CrossRef]

13. Casanovas, R.; Prieto, E.; Salvadó, M. Calculation of the ambient dose equivalent H*(10) from gamma-ray spectra obtained with scintillation detectors. *Appl. Radiat. Isot.* **2016**, *118*, 154–159. [CrossRef]

14. Buzhan, P.; Karakash, A.; Teverovskiy, Y. Silicon Photomultiplier and CsI(Tl) scintillator in application to portable H*(10) dosimeter. *Nucl. Instrum. Methods Phys. Res. Sect. A Accel. Spectrometers Detect. Assoc. Equip.* **2018**, *912*, 245–247. [CrossRef]

15. Camp, A.; Vargas, A. Ambient dose estimation H*(10) from LaBr$_3$(Ce) spectra. *Radiat. Prot. Dosim.* **2013**, *160*, 264–268. [CrossRef] [PubMed]

16. Petoussi-Henss, N.; Bolch, W.; Eckerman, K.; Endo, A.; Hertel, N.; Hunt, J.; Pelliccioni, M.; Schlattl, H.; Zankl, M. Conversion Coefficients for Radiological Protection Quantities for External Radiation Exposures. *Ann. ICRP* **2010**, *40*, 1–257. [CrossRef]

17. Saito, K.; Petoussi-Henss, N.; Zankl, M. Calculation of the Effective Dose and Its Variation from Environmental Gamma Ray Sources. *Heal. Phys.* **1998**, *74*, 698–706. [CrossRef] [PubMed]

18. ICRU. Endo on behalf of ICRU Report Committee 26 on Operational Radiation Protection Quantities for External Radiation Operational quantities and new approach by ICRU. *Ann. ICRP* **2016**, *45*, 178–187. [CrossRef]

19. Schulz, E.; Speekenbrink, M.; Krause, A. A tutorial on Gaussian process regression: Modelling, exploring, and exploiting functions. *J. Math. Psychol.* **2018**, *85*, 1–16. [CrossRef]

20. Ebden, M. Gaussian Processes: A Quick Introduction. *arXiv*, 2015; arXiv:1505.02965. Available online: https://arxiv.org/pdf/1505.02965.pdf(accessed on 19 May 2020).

21. Goorley, J.T.; James, M.R.; Booth, T.E.; Brown, F.B.; Bull, J.S.; Cox, L.J.; Durkee, J.W.; Elson, J.S.; Fensin, M.L.; Forster, R.A.; et al. *MCNP6 User's Manual, Version 1.0*; Los Alamos National Laboratory: Los Alamos, NM, USA, 2013.

22. Shi, H.-X.; Chen, B.-X.; Li, T.-Z.; Yun, D. Precise Monte Carlo simulation of gamma-ray response functions for an NaI(Tl) detector. *Appl. Radiat. Isot.* **2002**, *57*, 517–524. [CrossRef]

23. Jeon, B.; Kim, J.; Moon, M.; Cho, G. Parametric optimization for energy calibration and gamma response function of plastic scintillation detectors using a genetic algorithm. *Nucl. Instrum. Methods Phys. Res. Sect. A Accel. Spectrometers Detect. Assoc. Equip.* **2019**, *930*, 8–14. [CrossRef]

24. Ji, Y.-Y.; Chang, H.-S.; Lim, T.; Lee, W. Application of a SrI$_2$(Eu) scintillation detector to in situ gamma-ray spectrometry in the environment. *Radiat. Meas.* **2019**, *122*, 67–72. [CrossRef]

Article

Radioisotope Identification and Nonintrusive Depth Estimation of Localized Low-Level Radioactive Contaminants Using Bayesian Inference

Jinhwan Kim [†], Kyung Taek Lim [†], Kilyoung Ko, Eunbie Ko and Gyuseong Cho *

Department of Nuclear & Quantum Engineering, Korea Advanced Institute of Science and Technology, 291 Daehak-ro, Yuseong-gu, Daejeon 34141, Korea; kjhwan0205@kaist.ac.kr (J.K.); kl2548@kaist.ac.kr (K.T.L.); coltom@kaist.ac.kr (K.K.); cutsky@kaist.ac.kr (E.K.)
* Correspondence: gscho@kaist.ac.kr
† Both authors contributed equally to this work.

Received: 28 November 2019; Accepted: 19 December 2019; Published: 23 December 2019

Abstract: Obtaining the in-depth information of radioactive contaminants is crucial for determining the most cost-effective decommissioning strategy. The main limitations of a burial depth analysis lie in the assumptions that foreknowledge of buried radioisotopes present at the site is always available and that only a single radioisotope is present. We present an advanced depth estimation method using Bayesian inference, which does not rely on those assumptions. Thus, we identified low-level radioactive contaminants buried in a substance and then estimated their depths and activities. To evaluate the performance of the proposed method, several spectra were obtained using a 3×3 inch hand-held NaI (Tl) detector exposed to Cs-137, Co-60, Na-22, Am-241, Eu-152, and Eu-154 sources (less than 1μCi) that were buried in a sandbox at depths of up to 15 cm. The experimental results showed that this method is capable of correctly detecting not only a single but also multiple radioisotopes that are buried in sand. Furthermore, it can provide a good approximation of the burial depth and activity of the identified sources in terms of the mean and 95% credible interval in a single measurement. Lastly, we demonstrate that the proposed technique is rarely susceptible to short acquisition time and gain-shift effects.

Keywords: remote depth profiling; radioisotope identification; Bayesian inference; uncertainty estimation; gamma spectral analysis; low-level radioactive contaminants; nuclear decommissioning; low-resolution detector

---

1. Introduction

Sites near nuclear power plants are susceptible to large-scale land and building contamination because of the significant amount of radioactive waste generated by such facilities. It is, therefore, important to acquire information on the wastes present on these sites on behalf of project management and engineering services working on environmental restoration [1–3]. In particular, depth profiling of radioactive contaminants is critical for determining the most cost-effective decommissioning strategy, because the quantity of radioactive waste required for disposal can be reduced considerably by removing surface contamination at varying depths [4]. Nevertheless, the task of depth profiling is still difficult to achieve because porous materials such as soil and concrete covering the contaminants can act as a shield, resulting in the attenuation of emitted radiation.

One example of such waste is on the beaches of Dounreay in Northern Scotland, where radioactive soil contaminants are widely spread along the beach [5,6]. This is due to the so-called Dounreay hot particles that are mainly composed of Cs-137 and Co-60, released from the fuel processing of the Material Test Reactor at the Dounreay nuclear facility. Other examples of buried radioactive contaminants include orphan radioactive sources [7]. An orphan source is generally a sealed source of radioactive material that has been lost, abandoned, misplaced, stolen, or otherwise transferred without proper authorization [8].

Therefore, various non-destructive methods for remote-depth profiling have been reported in many papers [9–18]. However, the majority of the nonintrusive methods reported in these studies were based on a frequentist approach; that is, they required repeated measurements in order to provide a mean value with a standard error. Also, the maximum detectable depth of these methods was not sufficient to detect deeply buried contaminants. Therefore, a new approach, based on Bayesian inference, has recently been developed [19] to overcome the limitations imposed by older methods. This method can offer more reliable results because the output of burial depth analysis can be expressed as a probability distribution, even in a single measurement. In addition, its capability for maximum detectable depth for weak activity of the 0.94-μCi Cs-137 and 0.69-μCi Co-60 sources is superior in comparison with the existing methods. However, this method still assumes that only a single radioisotope is present in the substance and that no other radioisotopes will interfere with the measurement; a common assumption that is prevalent in other studies. But such assumptions can seriously undermine the results of a burial depth analysis in which there are different or multiple radioisotopes present.

Consequently, the objective of this study is first to identify all low-level radioactive contaminants buried in any substance, and then estimate remote depth profiling for the identified radioisotopes using Bayesian inference. In this study, radioactive sources of Cs-137, Co-60, Na-22, Am-241, Eu-152, and Eu-154, which are common elements encountered during decommissioning of nuclear facilities, were considered for the depth profiling. For convenience, the set of these radioisotopes will hereafter be referred to as the radioisotope library. Experimental results analyzed from various spectra, composed of not only single but also multiple radioisotopes, have been addressed to evaluate the performance of the proposed method. Furthermore, we have investigated the depth estimation performance of the proposed method in terms of data acquisition time and gain-shift effects due to calibration drift.

2. Materials and Methods

2.1. Bayesian Inference

Probability is one of the quantities that measure an event with an uncertainty that is associated with that particular event. There are two general philosophies providing different interpretations of probability: namely, frequentist inference and Bayesian inference [20]. In a frequentist approach, the probability is associated with the long-term frequency or proportion of events, in which the unknown parameters are treated as fixed values. That is, a frequentist does not associate probabilities with random variates, and only repeatable events can have probabilities in a statistical process. In contrast, a Bayesian approach is rooted in the belief that probabilities can be associated with unknown parameters (i.e., treated as random variables) to represent the uncertainty in any occurrence. That is, it can lead to much more intuitive results. For example, suppose you want to know the possibility that Korea will host the next World Cup. Bayesians are willing to assign a legitimate probability to Korea hosting the next World Cup based on the degrees of belief on the possible outcomes and every available information. Unlike Bayesians, frequentists do not assign any numerical probability to the same event because the World Cup cannot be regarded as a hypothetically repeatable process. This is a philosophical issue that frequentists can run into [21]. Also, some of the resultant interpretations are not particularly intuitive.

A Bayesian inference determines the probability distribution over the parameter or equivalently, the posterior distribution $p(\theta|y)$ of random variables θ, given prior distributions $p(\theta)$, and likelihood function $p(y|\theta)$ by applying Bayes' theorem:

$$p(\theta|y) = \frac{p(y|\theta)\,p(\theta)}{p(y)},\tag{1}$$

In the past, the challenge of applying the Bayesian inference to real-field applications was mainly around the computation requirement for the intractable high-dimensional integrals in the evidence $p(y)$. However, it is now possible, owing to recent advances made in computation technology and in marginal-estimation techniques. The Markov Chain Monte Carlo (MCMC) algorithm is a technique that is widely used for approximate inference, in which the posterior distribution is estimated through a collection of samples via the Markov process [22]. Since the late 1940s, there has been tremendous progress in the field of statistics, seeing the development of such techniques as the Metropolis Hasting algorithm, the Hamiltonian Monte Carlo, and more recently, the No-U-Turn sampler [23]. These algorithms were based on MCMC so that they could obtain the posterior probability of parameters with accuracy. However, their relatively high costs in computation and their inefficient processes have hindered their usage in real-world applications. An alternative method that can overcome these limitations is to convert the computation of $p(\theta|y)$ to an optimization problem, also known as variational inference.

With variational inference, we assume there is a parameterized family of distributions $q(\theta; v)$ (or equivalently, a variational distribution); then, we find the setting of the parameters that minimize the Kullback-Leibler (KL) divergence to the posterior distribution of interest:

$$v^* = argminKL\big(q(\theta; v)\big\|p(\theta|y)\big).\tag{2}$$

The optimized $q(\theta; v^*)$ is then regarded as an approximation to the posterior distribution. Since the KL divergence involving the posterior distributions lacks an analytic form, we instead maximize the evidence lower bound (ELBO):

$$\mathcal{L}(v) = \mathbb{E}_q[logp(\theta, y)] - \mathbb{E}_q[logq(\theta; v)].\tag{3}$$

This can be simplified further by taking a mean-field approximation, where the parameters in the variational family are assumed to be fully factorized to independent variables. However, the difficulties arising from the model-specific derivations and implementations in developing such algorithms still hinder its use in practical applications. However, automatic differentiation variational inference (ADVI), which is a gradient-based method, can resolve this complexity in computation by providing a recipe for an automatic solution based on variational inference [24]. The underlying idea of ADVI is to transform the space of latent variables and to automate derivatives of the joint distribution by relying on the capabilities of probabilistic programming systems. For programming ADVI computation, we used Python language with the probabilistic programming framework of PyMC3 to establish a probability model and execute variational inference.

2.2. Model Specification

By defining a mathematical model that describes an observed spectrum in terms of the burial depth, activity, and shift degree of the spectrum, we can identify buried radioisotopes and obtain the posterior distribution of the depth and activity of the identified sources. Such a model can be established by extending the model defined by Kim et al. [19,25]

$$M_i = \sum_{j=1}^{J} \frac{A_j P_j \delta_j}{4\pi h^2} e^{-\mu_A h} f\big(z_j,\ \eta i\big) + cB_i \ for \ i = 1, \ldots, K\tag{4}$$

Here, M_i is the measured spectrum (s^{-1}) with i representing the channel ($0 < i \leq K$); J is the total number of radioisotopes; A_j is the activity of the radioisotope (µCi); P_j is the total sum of gamma emission probabilities within the energy range of interest (i.e., 2.8 γs^{-1}Bq^{-1} for the 511 and 1275 keV gamma rays of Na-22); μ_A is the linear attenuation coefficient of gamma-ray in air (cm^{-1}); h is the detection height measured from the detector to the surface of a given material (cm); z is the buried depth of a radioactive source ($0 \leq z \leq D$ cm) in a material from the front surface; η is the shift degree of the spectrum; B_i is the background spectrum measured for K channels with c being its proportionality constant; δ is the effective front area (cm^{-1}); and $f(z, \eta i)$ is the bilinear interpolation function. Computation of $f(z, \eta i)$ requires a spectrum measurement at depths ranging from 0 to D cm at certain intervals to determine the $K \times D$ response matrix for a radioisotope. Consequently, $f(z, \eta i)$ can be interpolated using the closest points to the $f(z, \eta i)$ among the known values of depths and channels from the $K \times D$ response matrix [19]. The parameter δ can be obtained experimentally by placing a source on the material surface (that is, at 0 cm depth), which can be expressed as:

$$\delta = \frac{4\pi r^2 N}{APe^{-\mu_A r}},$$

(5)

where N is the total net counts of the spectrum (s^{-1}), and r is the detection height (cm) between the detector and the surface of a material.

Thus, the proposed model defines the function $f(\mathbf{z}, \mathbf{A},\ \eta,\ c)$ where the variable marked in bold type represents a vector notation. In practice, the existence of inevitable uncertainties inherent to the physical processes, such as radioactive disintegration, has an effect on the measured spectrum. In this regard, we can assume that the spectrum is normally distributed with a zero mean and variance of σ^2

$$P(M|\mathbf{z}, \mathbf{A},\ \eta,\ c) = N\big(f(\mathbf{z}, \mathbf{A},\ \eta,\ c),\ \sigma^2\big).$$

(6)

The availability of prior distributions for $\mathbf{z},\ \mathbf{A},\ \eta,\ c,$ and, σ^2, which represents our knowledge of the parameters before taking any measurements, is assumed by Kim et al. [19]. That is, $\mathbf{A}, c,$ and σ^2 followed gamma distributions with parameters (1, 1); z and η followed uniform distributions with parameters (0, 18) and (0.85, 1.15), respectively. These prior distributions reflected our belief that the sources might be buried less than 18 cm in the sand, and their activities would be low.

2.3. Procedures on Spectral Analysis

The spectral analysis of the depth estimation is a two-step process. First, the radioisotopes that are least likely to have generated an observed spectrum are excluded according to certain criteria [25]. This step is necessary because the model assigns a probability distribution to the parameters of every radioisotope present in the radioisotope library. For instance, the ratio of the standard deviation σ_j to the mean u_j of a radioisotope, i.e., relative standard deviation (RSD), can have a large value where a certain radioisotope in the library is not contributing to the spectrum. In terms of the magnitude of RSD, a small value suggests that the data are clustered tightly around the mean while the opposite is true in a large value of RSD. In addition, a radioisotope that does not attribute to the spectrum may have a relatively negligible contribution. The relative contribution (RC) of the radioisotope, C_j to the spectrum can be expressed as:

$$C_j = \frac{\dfrac{A_j P_j}{z_j^2}}{\sum_{j=1}^{J} \dfrac{A_j P_j}{z_j^2}}.$$

(7)

Here, radioisotopes can be regarded as present when the following conditions are satisfied:

$$C_j > 3\% \text{ and } \frac{\sigma_i}{u_i} < 0.2.$$

(8)

These thresholds are subject to change depending on the situation. The first analysis can be thought as the identification step. Second, the identified radioisotopes through the first analysis are analyzed to obtain their final depths and activities.

2.4. Experimental Setup

Gamma-ray spectra were obtained on radioactive sources buried in a sandbox filled with fine silica sand by using a 3 × 3 inch hand-held NaI (Tl) detector (NUCARE, Rad IQ™ HH200, Incheon, Korea) that was located 3 cm away from the surface of the box, as depicted in Figure 1a. The detector was used only for the purpose of acquiring gamma spectra and the recorded raw data were then processed and analyzed separately through Python. The sandbox was composed of 0.3 cm-thick acrylic sheet forming a tank of 50 cm × 40 cm × 40 cm (length × width × height). The thickness of the acrylic sheets was chosen so that the gamma rays emitting from the source would be scattered in the sand matrix. The activities of the sources used for the experiments were 0.94 µCi, 0.69 µCi, 0.50 µCi, 0.90 µCi, 0.89 µCi, and 0.84 µCi for Cs-137, Co-60, Na-22 Am-241, Eu-152, and Eu-154, respectively. In addition, the sources were buried in a graduated box (50 cm × 0.3 cm × 0.3 cm) that was inserted into the main box to position the sources at the exact location in relation to the front of the sandbox surface, as shown in Figure 1b.

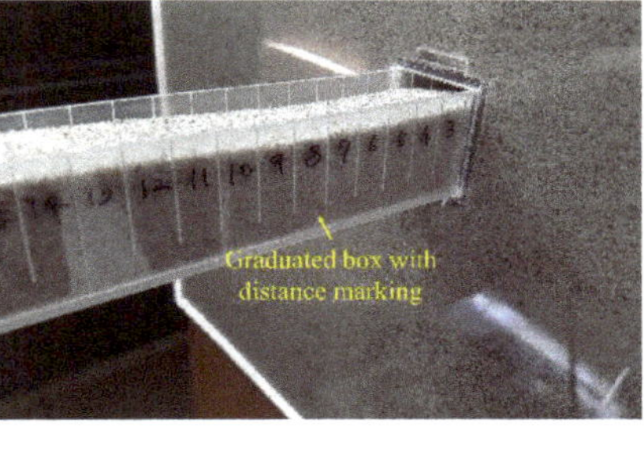

(a)   (b)

Figure 1. (**a**) An acrylic box filled with sand and an NaI (Tl) detector for gamma spectroscopy; and (**b**) a graduated box marked with the buried distance of the source measured from the front surface of the sandbox.

The response matrix was obtained by placing the sources in the graduated sandbox at depths of 0 cm, 3 cm, 7 cm, 12 cm, and 18 cm. Then, the spectra were measured at each depth for 30 min to ensure that minimal statistical fluctuation was achieved. A background spectrum for the response matrix was obtained under identical conditions in the absence of sources. For the energy range of spectra, values were chosen from 20 to 1600 keV (i.e., 563 channels). During these experiments, energy calibration was performed prior to taking each measurement via the built-in automatic calibration function provided by the detector system. This function is based on the energy emitted by the primordial radioisotope of K-40 (1461 keV). The automatic calibration function was not used for the acquisition of the test spectra because this method automatically compensates the gain-shift effects because of changes in ambient temperature or calibration drifts [19].

3. Results

3.1. Case 1: Single Radioisotope

Figure 2 shows the joint probability distributions between the depth and activity of the radioisotopes analyzed for a spectrum, measured for 300 s for a Cs-137 source buried at a depth of 3 cm. From the first analysis, we can clearly see that the distribution of the Cs-137 is clustered tightly around the mean, while the distribution of the other radioisotopes (i.e., Co-60, Na-22, Am-241, Eu-152, and Eu-154) is spread along high values of the depth at low activity. As shown in Figure 3, the values of the RCs and RSDs for the five radioisotopes did not satisfy the criteria mentioned in Section 2.3, and therefore only the radioisotope of Cs-137 provided any notable contribution to the spectrum. The second analysis was then performed on the Cs-137 to determine its burial depth and activity. The result confirmed that the joint probability distribution of the Cs-137 was closely centered around the true value of the depth and activity (i.e., 3 cm and 0.942 µCi). It is not always true, however, that the distributed results of the first and second analyses will yield nearly the same output, as we have seen on this occasion. This is because it is possible for this method to induce a distortion in the analysis results by assigning a biased mean of activity to certain radioisotopes during the identification step [25].

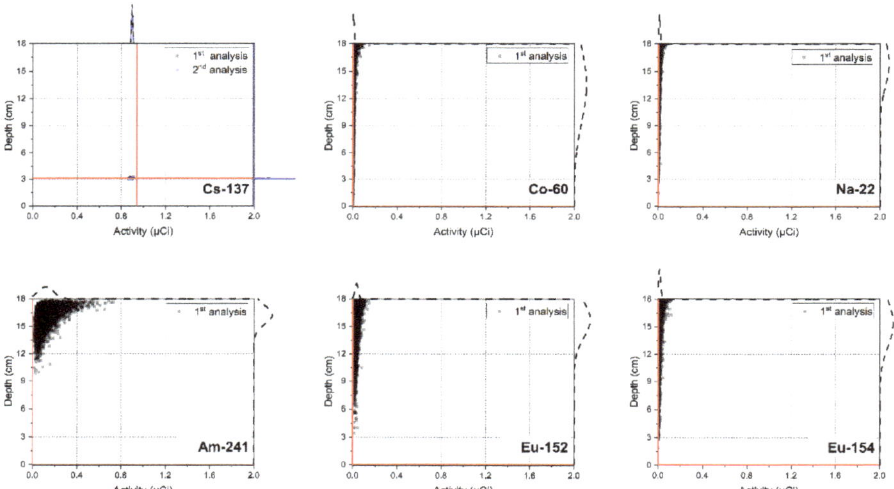

Figure 2. Joint distributions between the depth and activity of the radioisotopes in the radioisotope library for a spectrum acquired for 300 s with a Cs-137 source buried in sand at the depth of 3 cm. The scatter dots represent the correlations between the depth and activity, while red lines and the curves outside the plot area represent their true values and corresponding densities, respectively. The distribution from the first analysis is obscured by that of the second analysis and is hardly visible in the plot for Cs-137.

Figure 4 shows the estimated depth and activity with a 95% credible interval for all single radioisotopes, namely: Cs-137, Co-60, Na-22, Am-241, Eu-152, and Eu-154, buried in sand over a range of 0–15 cm at intervals of 3 cm. The spectra for the analysis were measured for 300 s. From the experimental results, we found that the proposed technique was capable of correctly identifying the buried radioisotopes and determining the depth of the identified radioisotopes with the exception of the Am-241 source at burial depths exceeding 9 cm. At these depths, RSD and RC values for all radioisotopes in the radioisotope library did not meet the criteria, meaning that there were no other radioisotopes affecting the spectra except for the background radiation. This was mainly due to the high attenuation of low-energy photons (e.g., 59 keV) emitted by Am-241. As a consequence, the spectra obtained with the Am-241 source buried deeply became almost indistinguishable from a background spectrum, as shown in Figure 5. Excluding the Am-241, the results confirmed that the true depth was approximated by the mean value of the estimated depth with a 95% credible interval for all radioisotopes with very weak activities that were buried in sand over a range of 0–15 cm; the estimated depths at a depth of 6 cm tend to be slightly higher, probably because of the discrepancy between the measured spectra and the spectra calculated by interpolation. In addition, the estimated mean values of the activity with a 95% credible interval for the identified radioisotopes were in close agreement with the true values. Likewise, the trend in the relationship between the depth and the activity was also in agreement with the results report by Kim et al. [19].

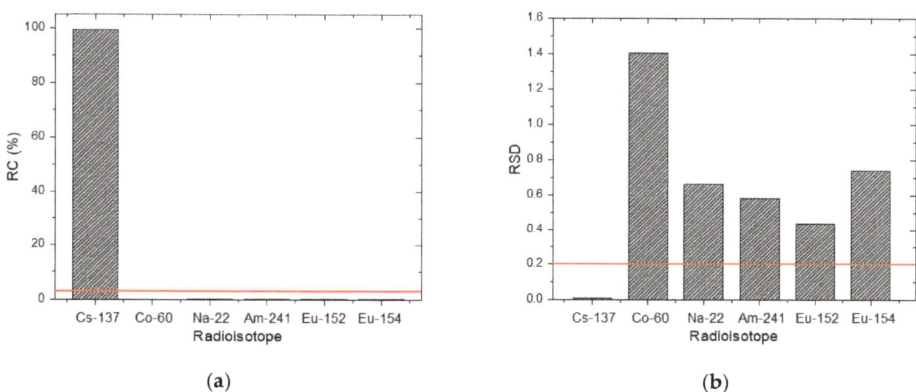

(a)   (b)

Figure 3. (a) RC and (b) RSD of the radioisotopes in the radioisotope library for a spectrum acquired for 300 s with a Cs-137 source buried in sand at a depth of 3 cm. The red lines denote criteria for the RC and RSD (i.e., 3% and 0.2, respectively).

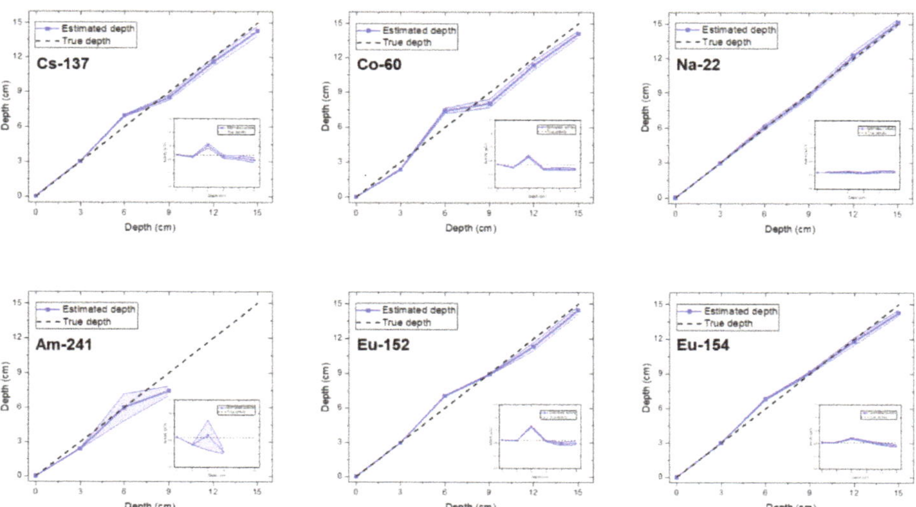

Figure 4. Estimated depth and activity with a 95% credible interval for the single radioisotope of Cs-137, Co-60, Na-22, Am-241, Eu-152, and Eu-154 buried in sand over the range of 0–15 cm at 3 cm intervals. The inset shows the estimated activity of the corresponding radioisotope in each figure.

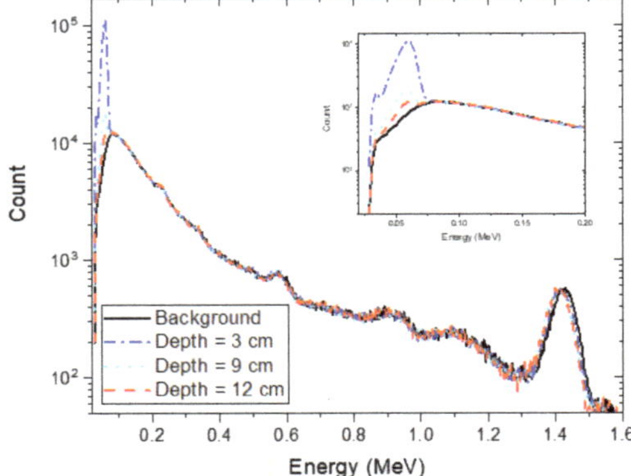

Figure 5. Experimental spectra with an acquisition time of 30 min for an Am-241 source buried in sand at depths of 3 cm (blue dash-dotted line), 9 cm (sky-blue dotted line) and 12 cm (red dashed line). The black solid line represents the background spectrum with the same acquisition time. The inset shows the enlargement of the spectra in the low-energy region.

3.2. Case 2: Multiple Radioisotopes and Data Acqusition Time

To validate the performance of the proposed method in cases where multiple radioisotopes were buried at different depths, we measured the spectra with different acquisition times of 10 s, 30 s, and 300 s for Eu-152 and Eu-154 sources that were buried in sand at 3 cm and 6 cm in depth, respectively, as shown in Figure 6. The reason for the acquisition of the spectra with the reduced acquisition times was to verify the performance of the proposed method in large-scale field measurements that require a rapid acquisition. As shown in Figure 7, the radioisotopes that had not contributed to the spectra (i.e., Cs-137, Co-60, Na-22, and Am-241) could be rejected in the identification step, meaning that our method can detect the correct radioisotopes for the spectra, even with short acquisition times. The estimated depth and activity of the identified radioisotopes for the spectra are illustrated as joint probability distributions in Figure 8. From these results, the distributions between the depth and activity for the radioisotopes were more closely clustered with increasing acquisition time. In addition, the center of the distributions got closer to the true values. This can be more clearly seen in Figure 9 where the error bar is in the form of mean and 1.96 standard error (or equivalently, a 95% credible interval). This shows that the true values of the depth for the identified radioisotopes indeed fell within the 95% credible interval of the estimated depths analyzed for the spectra. To our surprise, the mean values of the depth analyzed even for the acquisition time of 10 s closely agreed with the true values, which is a much better result than that reported by Kim et al. [19]. This was due to the use of the larger-size more efficient detector. The estimated values of the activity for Eu-154 deviated slightly from the true values. This was possibly due to a position error in the sources during the measurements. In this proposed method, the determination of depth depends primarily on a spectral shape; that is, our approach determines the burial depths of radioisotopes that are most likely to have produced an observed spectrum through the combination of spectral shapes of each radioisotope at varying depths. In contrast, the activity was calculated based mainly on the determined depth and counts in the observed spectrum (see Equation (4)). In this regard, the activity should be accurately estimated once the depths are exactly estimated and the acquisition time is sufficient to reduce the statistical fluctuation present in the observed spectrum.

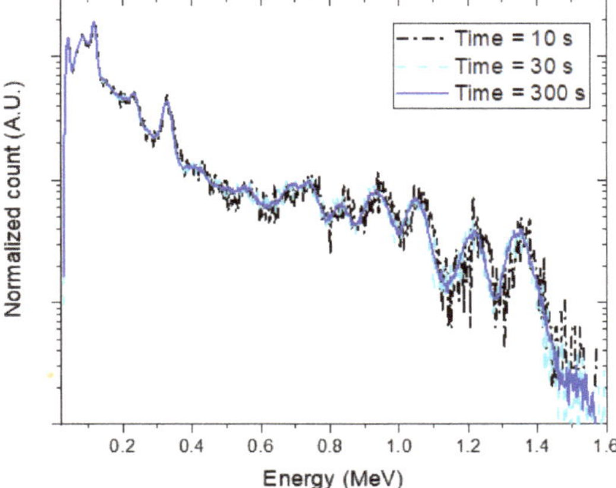

Figure 6. Experimental spectra with different acquisition times of 10 s (black dash-dotted line), 30 s (sky-blue dashed line) and 300 s (blue solid line) for the following radioisotopes buried in sand: Eu-152, 3 cm and Eu-154, 6 cm. Obtained spectra were normalized to the total count over the energies of interest for comparison.

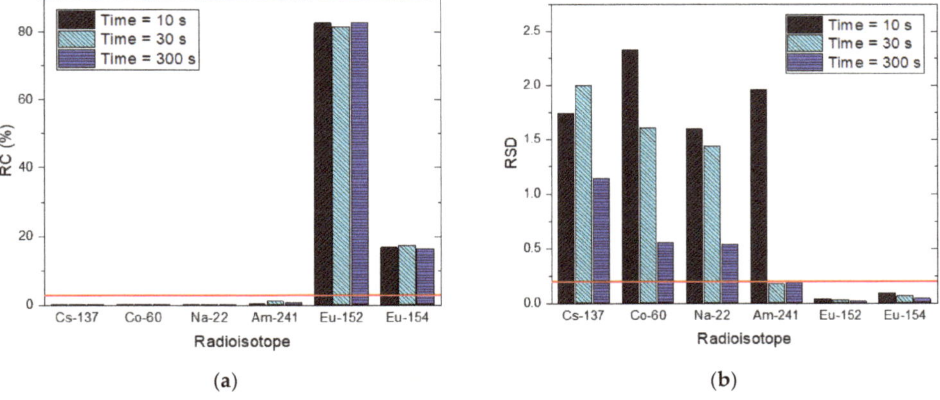

Figure 7. (a) RC and (b) RSD of radioisotopes in the radioisotope library for three spectra with acquisition times of 10 s, 30 s and 300 s with the following radioisotopes buried in sand: Eu-152, 3 cm and Eu-154, 6 cm. The red lines denote criteria for the RC and RSD (i.e., 3% and 0.2, respectively).

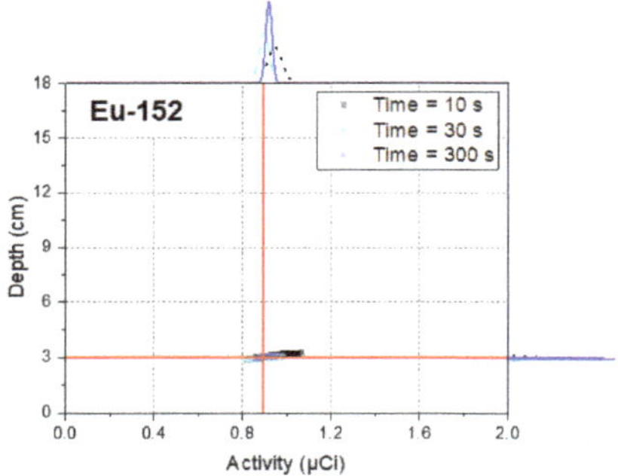
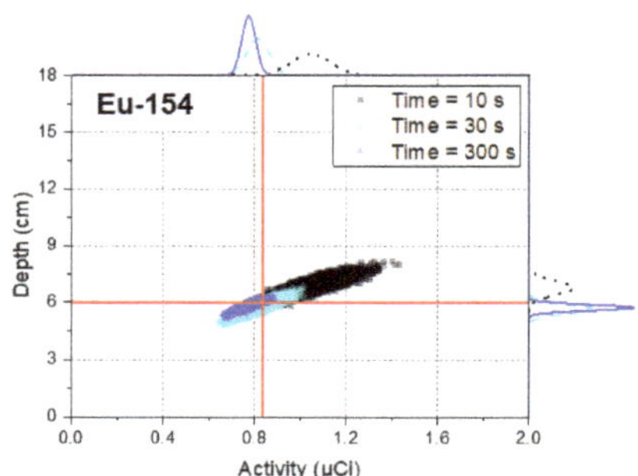

Figure 8. Joint distributions between the depth and activity of identified radioisotopes (i.e., Eu-152 and Eu-154) for experimental spectra acquired with 10 s, 30 s, and 300 s for Eu-152 and Eu-154 sources buried in sand at depths of 3 cm and 6 cm, respectively. The scatter dots represent the correlations between the depth and activity, while the red lines and the curves outside the plot area represent their true values and corresponding densities, respectively.

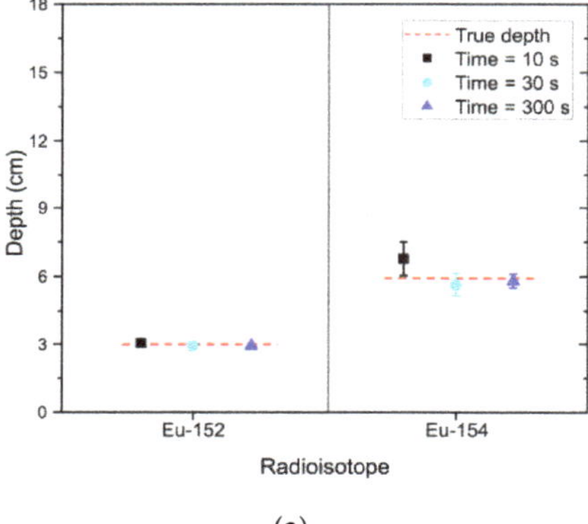
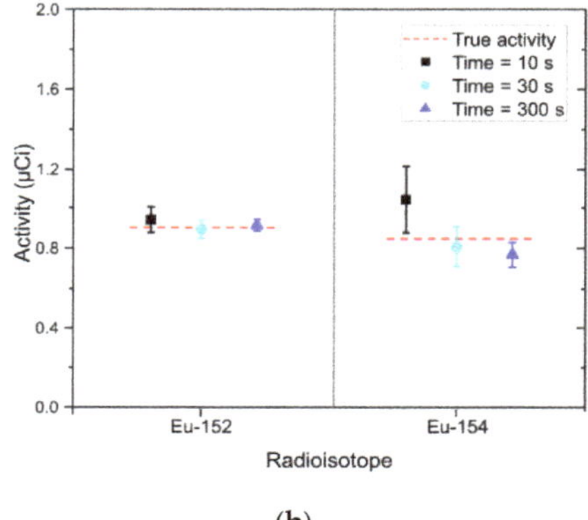

(a) (b)

Figure 9. (a) Estimated depth and (b) activity of identified radioisotopes (i.e., Eu-152 and Eu-154) in the form of mean and 1.96 standard error analyzed for experimental spectra with acquisition times of 10 s (black square), 30 s (sky-blue circle), and 300 s (blue triangle). The spectra were acquired with Eu-152 and Eu-154 sources buried in sand at depths of 3 cm and 6 cm, respectively.

Figure 10 shows more complex spectra obtained with acquisition times of 10 s, 30 s, and 300 s for Na-22, Am-241, and Eu-152 with burial depths of 10 cm, 3 cm, and 8 cm, respectively. Similar to the previous case, the radioisotopes that are not part of the spectra (i.e., Cs-137, Co-60, and Eu-154) were rejected in the identification step and we could therefore correctly detect the radioisotopes of Na-22, Am-241, and Eu-152. The estimated depth and activity of the identified radioisotopes were reported in terms of mean and 1.96 standard errors, as shown in Figure 11. Here, trends relating to the acquisition time can be observed on the standard errors that are similar to those reported in Figure 9. In particular, the estimated depth of Na-22 for the spectrum with the acquisition time of 10 s showed a relatively large standard error. Also, it deviated from the true value because of the statistical fluctuations in the spectrum. Except this, the results indicated that there was a satisfactory agreement between the estimated and true values for the complex spectrum even where there were the short acquisition times. However, the estimated activities showed relatively more significant deviations from the true values because the activity is inversely related to the square of the depth (see Equation (4)).

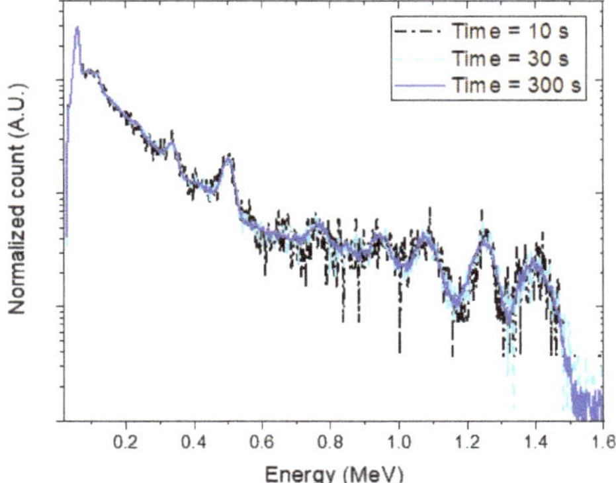

Figure 10. Experimental spectra with different acquisition times of 10 s (black dash-dotted line), 30 s (sky-blue dashed line), and 300 s (blue solid line) for the following radioisotopes buried in sand: Na-22, 10 cm; Am-241, 3 cm; and Eu-152, 8 cm. Obtained spectra were normalized to the total count over the energies of interest for comparison.

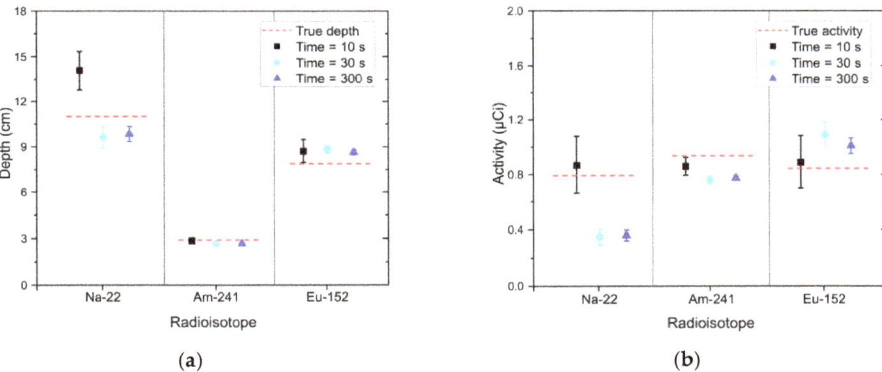

Figure 11. (**a**) Estimated depth and (**b**) activity of identified radioisotopes (i.e., Na-22, Am-241, and Eu-152) in the form of mean and 1.96 standard error analyzed for experimental spectra with acquisition times of 10 s (black square), 30 s (sky-blue circle), and 300 s (blue triangle). The spectra were acquired with Na-22, Am-241, and Eu-152 sources buried in sand at depths of 11 cm, 3 cm, and 8 cm, respectively. The red dotted lines denote the true values.

3.3. Effect of Gain Shift

To investigate how well the proposed technique would accurately analyze shifted spectra because of gain-shift effects, we acquired spectra at an acquisition time of 30 s for the Eu-152 and Eu-154 sources at burial depths of 3 cm and 6 cm after calibration drifts had occurred in the detector. Figure 12 shows these spectra with two different magnitudes of the shift (blue-sky dotted line and blue solid line referred to as "G1" and "G2") and the normal spectrum (black dash-dotted line) for comparison. In fact, these shifted spectra would be very difficult to analyze without proper recalibration settings; the position of the original photo-peak in the high-energy region of the normal spectrum was overlapped completely by another peak in the G2 spectrum. Nevertheless, the proposed method was able to exclude the radioisotopes that had not contributed to the shifted spectra in the identification step. For the G1 spectrum, the estimated depths of Eu-152 and Eu-154 closely agree with the true values as shown in Figure 13a. For the G2 spectrum, the mean value of the estimated depth for Eu-154 was

7.52 ± 0.38 (1.96σ) cm, which was slightly overestimated. The activity tended to be underestimated against the determined depths as the spectrum shifted in a positive direction (see Figure 13b). This was possibly due to an increase in full width at half maximum as the spectrum moves to higher energies, resulting in a reduction in the maximum counts in the region of the photo-peaks, which does not occur in spectra that were shifted mathematically via interpolation. Overall, the presented results demonstrated a capability to accommodate a shift in the spectra caused by calibration drift in the complex spectra of multiple radioisotopes.

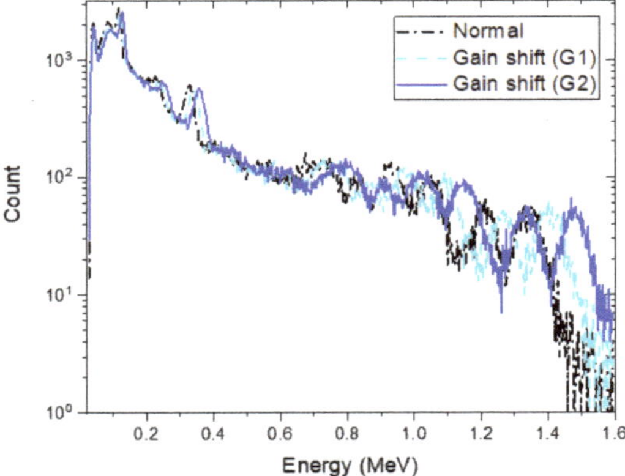

Figure 12. Experimental spectrum composed of Eu-152 and Eu-154 sources buried in sand at depths of 3 cm and 6 cm, respectively (black das-dotted line) and its shifted spectra (sky-blue dashed line and blue solid line) having different magnitudes of the shift due to calibration drifts. The spectra were acquired for 30 s.

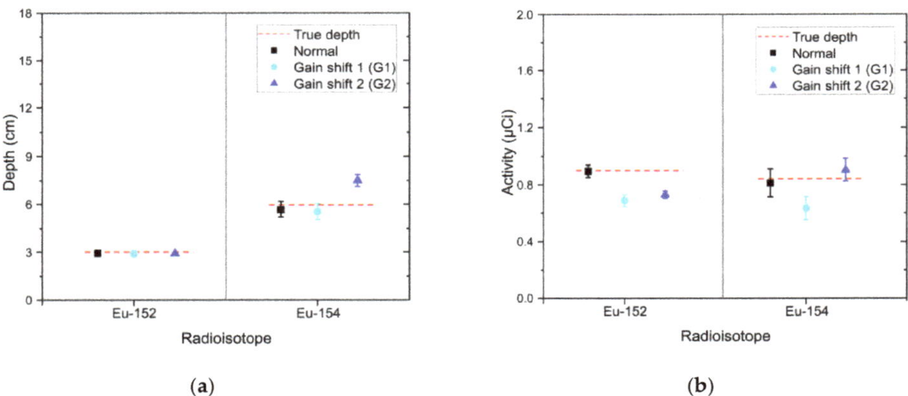

Figure 13. (a) Estimated depth and (b) activity of identified radioisotopes (i.e., Eu-152 and Eu-154) in the form of mean and 1.96 standard error analyzed for a 30-s measured spectrum composed of Eu-152 and Eu-154 buried in sand at depths of 3 cm and 6 cm (black square) and its shifted spectra (sky-blue circle and blue triangle). The red dotted lines denote the true values.

4. Discussion

In this work, we demonstrated the estimation of remote depth profiling for contaminated low-level radioactive materials that are composed of single or multiple radioisotopes by applying Bayesian inference. An earlier report had already shown that this approach is reliable and robust because it allows us to offer the mean and standard error for the estimated depth and activity from a single measurement [19]. Also, the reported experimental results demonstrated that significant improvements in earlier findings had been achieved. First, the proposed technique does not rely on an assumption that we have foreknowledge of a radioisotope present at the site, or that only a single radioisotope exists there. Instead, this method first identifies unknown radioisotopes and then determines the depth and activity of the identified source(s). Thus, we are not only able to identify individual radioisotopes for spectra composed of multiple radioisotopes, but also to provide a good approximation of each one's depth and activity. Second, the results showed that this method can be applied to both low-level buried wastes and all gamma-emitting radioisotopes, regardless of the intensity of the gamma-ray energy or the number of gamma rays emitted, given that photons are not fully attenuated in a substance and contribute to the spectra to some degree. Lastly, we demonstrated that this method is also capable of accommodating the gain-shift effects in spectra with multiple radioisotopes.

One of the challenges associated with the measurement point is that it is difficult to find an optimal position for the detector in relation to the location of the buried radioactive contaminants. An alternative solution could be to position the detector at the location with the maximum intensity of total count rate. However, multiple contaminants buried at different depths may not be vertically positioned. If they are located in that way, the error of the x-y position causes an error of the burial depth (z position). Further work must be conducted to resolve this issue so that a better approximation of localized radioactive wastes in three dimensions can be achieved.

5. Conclusions

In this work, we presented a novel method for the remote depth estimation of unknown radioactive contaminants using Bayesian inference. Experimental results confirmed that this method correctly identifies radioactive contaminants composed of multiple radioisotopes as well as a single radioisotope and provides good estimates of depths buried in sand for the identified isotopes in a single measurement. In addition, we demonstrated that short acquisition time and gain-shift effects did not significantly degrade the analysis results for spectra composed of multiple radioisotopes. These results showed significantly improved remote depth estimation capability in comparison with the existing methods. Therefore, the proposed method is capable of achieving a rapid nonintrusive localization of buried low-level multiple radioactive contaminants through in situ measurement.

Author Contributions: Conceptualization, J.K. and G.C.; methodology, J.K. and K.T.L.; software, J.K. and K.K.; validation, J.K and K.T.L.; formal analysis, J.K. and K.T.L.; investigation, J.K., K.T.L., and E.K.; writing—original draft preparation, J.K.; writing—review and editing, G.C. and K.T.L.; visualization, J.K. and K.K.; supervision, G.C. and K.T.L.; funding acquisition, G.C. All authors have read and agreed to the published version of the manuscript.

Funding: This work was supported by NRF (National Research Foundation of Korea) Grant funded by the Korean Government (NRF-2018-Global Ph.D. Fellowship Program).

Conflicts of Interest: The authors declare no conflict of interest.

References

1. Characterization of Radioactively Contaminated Sites for Remediation Purposes. Available online: https://www-pub.iaea.org/MTCD/publications/PDF/te_1017_prn.pdf (accessed on 20 December 2019).
2. Radiological Characterisation for Decommissioning of Nuclear Installations. Available online: https://www.oecd-nea.org/rwm/docs/2013/rwm-wpdd2013-2.pdf (accessed on 20 December 2019).
3. Multi-Agency Radiation Survey and Site Investigation Manual (MARSSIM). (NUREG-1575, Revision 1). Available online: https://www.nrc.gov/reading-rm/doc-collections/nuregs/staff/sr1575/r1/ (accessed on 20 December 2019).
4. Sullivan, P.O.; Nokhamzon, J.G.; Cantrel, E. Decontamination and Dismantling of Radioactive Concrete Structures. *NEA News* **2010**, *28*, 27–29.
5. Dennis, F.; Morgan, G.; Henderson, F. Dounreay Hot Particles: The Story so Far. *J. Radiol. Prot.* **2007**, *27*, A3–A11. [CrossRef] [PubMed]

6. Dounreay Particles Advisory Group. Available online: https://assets.publishing.service.gov.uk/government/uploads/system/uploads/attachment_data/file/696380/DPAG_3rd__Report_September_2006.pdf (accessed on 20 December 2019).

7. Popp, A.; Ardouin, C.; Alexander, M.; Blackley, R.; Murray, A. Improvement of a High Risk Category Source Buried in the Grounds of a Hospital in Cambodia. In Proceedings of the 3th International Congress of the International Radiation Protection Association (IRPA), Glasgow, UK, 14–18 May 2012; pp. 1–10.

8. Control of Orphan Sources and Other Radioactive Material in the Metal Recycling and Production Industries. Available online: https://www-pub.iaea.org/MTCD/Publications/PDF/Pub1509_web.pdf (accessed on 20 December 2019).

9. Maeda, K.; Sasaki, S.; Kumai, M.; Sato, I.; Suto, M.; Ohsaka, M.; Goto, T.; Sakai, H.; Chigira, T.; Murata, H. Distribution of Radioactive Nuclides of Boring Core Samples Extracted from Concrete Structures of Reactor Buildings in the Fukushima Daiichi Nuclear Power Plant. *J. Nucl. Sci. Technol.* **2014**, *51*, 1006–1023. [CrossRef]

10. Shippen, B.A.; Joyce, M.J. Extension of the Linear Depth Attenuation Method for the Radioactivity Depth Analysis Tool (RADPAT). *IEEE Trans. Nucl. Sci.* **2011**, *58*, 1145–1150. [CrossRef]

11. Shippen, A.; Joyce, M.J. Profiling the Depth of Caesium-137 Contamination in Concrete via a Relative Linear Attenuation Model. *Appl. Radiat. Isot.* **2010**, *68*, 631–634. [CrossRef] [PubMed]

12. Adams, J.C.; Joyce, M.J.; Mellor, M. The Advancement of a Technique Using Principal Component Analysis for the Non-Intrusive Depth Profiling of Radioactive Contamination. *IEEE Trans. Nucl. Sci.* **2012**, *59*, 1448–1452. [CrossRef]

13. Adams, J.C.; Joyce, M.J.; Mellor, M. Depth Profiling 137Cs and 60Co Non-Intrusively for a Suite of Industrial Shielding Materials and at Depths beyond 50mm. *Appl. Radiat. Isot.* **2012**, *70*, 1150–1153. [CrossRef] [PubMed]

14. Adams, J.C.; Mellor, M.; Joyce, M.J. Determination of the Depth of Localized Radioactive Contamination by 137Cs and 60Co in Sand with Principal Component Analysis. *Environ. Sci. Technol.* **2011**, *45*, 8262–8267. [CrossRef] [PubMed]

15. Ukaegbu, I.K.; Gamage, K.A.A. A Novel Method for Remote Depth Estimation of Buried Radioactive Contamination. *Sensors* **2018**, *18*, 507. [CrossRef] [PubMed]

16. Ukaegbu, I.K.; Gamage, K.A.A. A Model for Remote Depth Estimation of Buried Radioactive Wastes Using CdZnTe Detector. *Sensors* **2018**, *18*, 1612. [CrossRef] [PubMed]

17. Ukaegbu, I.K.; Gamage, K.A.A.; Aspinall, M.D. Nonintrusive Depth Estimation of Buried Radioactive Wastes Using Ground Penetrating Radar and a Gamma Ray Detector. *Remote Sens.* **2019**, *11*, 141. [CrossRef]

18. Adams, J.C.; Mellor, M.; Joyce, M.J. Depth Determination of Buried Caesium-137 and Cobalt-60 Sources Using Scatter Peak Data. *IEEE Trans. Nucl. Sci.* **2010**, *57*, 2752–2757. [CrossRef]

19. Kim, J.; Lim, K.T.; Park, K.; Cho, G. A Bayesian Approach for Remote Depth Estimation of Buried Low-Level Radioactive Waste with a NaI (Tl) Detector. *Sensors* **2019**, *19*, 5365. [CrossRef] [PubMed]

20. Wagenmakers, E.-J.; Lee, M.; Lodewyckx, T.; Iverson, G.J. Bayesian Versus Frequentist Inference. In *Bayesian Evaluation of Informative Hypotheses*; Springer: New York, NY, USA, 2008; pp. 181–207. [CrossRef]

21. Philosophy of Statistics. Available online: https://plato.stanford.edu/entries/statistics/#pagetopright (accessed on 20 December 2019).

22. Andrieu, C.; Freitas, N.; Doucet, A.; Jordan, M.I. An Introduction to MCMC for Machine Learning. *Mach. Learn.* **2003**, *50*, 5–43. [CrossRef]

23. Mullachery, V.; Khera, A.; Husain, A. Bayesian Neural Networks. *arXiv* **2018**, arXiv:1801.07710.

24. Kucukelbir, A.; Tran, D.; Gelman, A.; Blei, D.M. Automatic Differentiation Variational Inference. *J. Mach. Learn. Res.* **2017**, *18*, 430–474.

25. Kim, J.; Taek, K.; Kim, J.; Kim, Y.; Kim, H. Quantification and Uncertainty Analysis of Low-Resolution Gamma-Ray Spectrometry Using Bayesian Inference. *Nucl. Instrum. Methods Phys. Res. Sect. A Accel. Spectrom. Detect. Assoc. Equip.* **2019**, in press. [CrossRef]

Article

Compact Viscometer Prototype for Remote In Situ Analysis of Sludge

Tomas Fried [1,2]*, David Cheneler [1], Stephen D. Monk [1], C. James Taylor [1] and Jonathan M. Dodds [2]

[1] Engineering Department, Lancaster University, Lancaster LA1 4YW, UK
[2] National Nuclear Laboratory, Workington CA14 3YQ, UK
* Correspondence: t.fried@lancaster.ac.uk

Received: 6 June 2019; Accepted: 24 July 2019; Published: 26 July 2019

Abstract: On the Sellafield site there are several legacy storage tanks and silos containing sludge of uncertain properties. While there are efforts to determine the chemical and radiological properties of the sludge, to clean out and decommission these vessels, the physical properties need to be ascertained as well. Shear behaviour, density and temperature are the key parameters to be understood before decommissioning activities commence. However, limited access, the congested nature of the tanks and presence of radioactive, hazardous substances severely limit sampling and usage of sophisticated characterisation devices within these tanks and therefore, these properties remain uncertain. This paper describes the development of a cheap, compact, and robust device to analyse the rheological properties of sludge, without the need to extract materials from the site in order to be analysed. Analysis of a sludge test material has been performed to create a suitable benchmark material for the rheological measurements with the prototype. Development of the device is being undertaken with commercial off the shelf (COTS) components and modern rapid prototyping techniques. Using these techniques, an initial prototype for measuring shear parameters of sludge has been developed, using a micro-controller for remote control and data gathering. The device is also compact enough to fit through a 75 mm opening, maximising deployment capabilities.

Keywords: rheology; nuclear decommissioning; rapid prototyping

---

1. Introduction

The landscape of the nuclear industry is changing nuclear power plants, and as reprocessing facilities approach the end of their planned lifespan, decommissioning has become a new focal point for the nuclear industry. This is reflected in academic research as well, with research moving from reprocessing operations to decommissioning [1]. An example of a decommissioning facility in the United Kingdom is the Sellafield site. Once the home of the first commercial nuclear reactor producing electricity on an industrial scale, the nuclear fuel reprocessing capability is approaching the end of its life cycle with the Thermal Oxide Reprocessing Plant (Thorp) facility stopping shearing in 2018 and Magnox reprocessing will end in 2020 [2,3].

Decommissioning of nuclear facilities requires a post-operational cleanout of active and hazardous materials. Using the Sellafield site as an example, there are a number of legacy tanks, silos and other containers with nuclear waste, often in the form of suspension, that will need to be cleaned out as part of the decommissioning plans. However, to accomplish this, analysis of remnant material must be performed to determine and minimise the risks, waste packages and ascertain the appropriate processes. Due to the hazardous nature of nuclear materials, this often has to be done remotely. Remote, in situ characterisation is, therefore, one of the key aspects of nuclear decommissioning [4].

The current strategy for analysis of remnant materials mainly focuses on visual inspection, radiological measurements and chemical composition analysis. Radiological measurements vary from using scintillators in storage tanks to determine the activity of remnant materials, pipe crawling robots assessing the contamination of pipework or autonomous platforms for analysis of floor contamination [5–8]. Current academic research focuses on imaging systems and the deployment of these in challenging environments. Advances in chemical composition analysis in nuclear facilities have introduced laser-induced breakdown spectroscopy (LIBS) as one potential method of remote analysis [9,10]. Visual inspection aids with assessing structures, waste locations and remnant material quantities in locations inaccessible to human workers. Most recently, advances in 3D scanning technologies (such as Light Detection and Ranging, or LiDAR) have been the focal point of research for visualizing remote areas. Other areas of research interest focus on the deployment platforms for both traditional camera systems and LiDAR in remote areas, whether through limited openings or underwater [11].

Currently used methods to ascertain rheological parameters of materials are predominantly rotational and oscillation rheometry using various geometries, either with samples or on-line [12–14]. Squeeze flow and capillary rheometry is frequently used in industries other than nuclear and academic research has been exploring these areas, too [15–17]. However, all of these methods utilise large, expensive equipment and development focus on increasing the precision and measurement range of devices.

In situ rheology in hazardous environments is, however, an unexplored research area. Research has been limited to test materials, or it relies on collecting samples from storage containers [18–20]. Novel methods for rheological measurements that have been developed are focused on miniaturisation and small sample methods, such as quartz crystal microbalance, but these are still in early research stages [21].

This paper presents work undertaken to develop a compact, cheap and robust device for shear behaviour analysis and its initial validation in a non-active laboratory environment. The aims are to propose a novel prototype device; to manufacture this device using rapid prototyping methods and commercial off the shelf (COTS) components; to calibrate the new device using industrial standard materials and to validate its performance at a laboratory bench scale. Section 2 provides a summary of materials and conventional instruments that have been used for analysis and manufacturing. Section 3 provides a description of the designed device, the tests performed, and the results of these tests. Finally, the discussion and conclusions are presented in Section 4.

2. Materials and Methods

A Bohlin CVO100 (Malvern Panalytical Ltd, Malvern, UK) was used as the baseline, commercially available, benchtop rheometer for the development of the device. All measurements made with the commercial rheometer were logarithmically spaced shear rate ramps from 0.1 to 100 s^{-1} at ambient temperatures between 21.5 and 22.5 °C, with a DIN standard V25 vane and cup and C25 bob and cup configurations. All measurements consisted of five measurement runs, and the results presented are averages of those five runs.

A silicone viscosity standard oil (Paragon Scientific Ltd, Merseyside, UK, VIS-RT1K-600) has been used as the calibration material. It is an industry-standard material used for calibrating rheometers and viscometers.

Two test materials were chosen to validate the performance of the device when measuring particulate suspensions. The first material was titanium dioxide (anatase, Sigma-Aldrich, Louis, MO, USA, 248576). This is an example of a lower viscosity, non-Newtonian particulate suspension. The TiO_2 was supplied as a powder, which was dried and subsequently dispersed in de-ionized water at a volume fraction of $\phi = 0.44$. TiO_2 has been historically used as a very safe test material in the nuclear industry [20]. Although it has since been replaced with other test materials, it remains an appropriate test material for the early stages of device development.

The second material was zirconium-molybdate (ZM) suspended in 0.5 M nitric acid. The sample was synthesised by Johnson Matthey and supplied as a suspension. Some of the supernate was decanted, and a sample was dried to ascertain the volume fraction of the suspension. ZM is an example of a higher viscosity test material. It is the most novel simulant being used in the nuclear industry [22].

A Malvern Mastersizer 3000 (Malvern Panalytical Ltd, Malvern, UK) particle characterisation system with a wet dispersion unit was used to ascertain the particle size distribution (PSD) of the two test materials, as shown in Figure 1 Both materials were dispersed in de-ionized water (refractive index 1.33). TiO_2 was stirred at 1800 rpm and measured at approximately 3.7% laser obscuration, using a refractive index 2.493 and an absorption index 1. ZM was stirred at 2400 rpm and measured at approximately 30.7% laser obscuration, using a refractive index 1.19 and an absorption index 0.

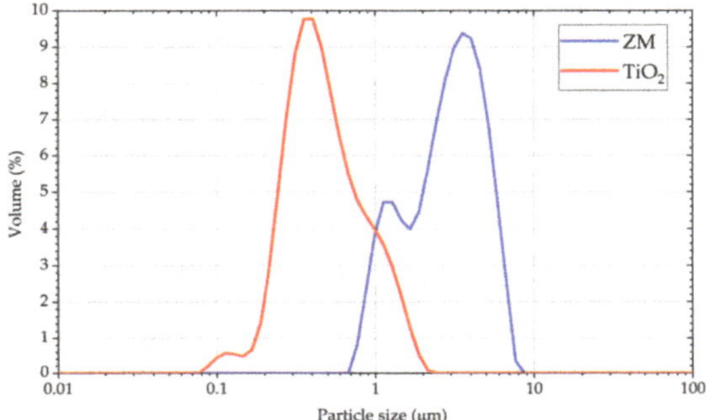

Figure 1. Particle size distribution (PSD) of the titanium dioxide and zirconium molybdate samples used in this research.

A majority of the structural components and other parts were manufactured using a fused deposition modelling 3D printer (Ultimaker 3 extended, Ultimaker, Utrecht, The Netherlands). Two different materials were utilised as follows. Linkages and outer casings not coming into contact with chemical substances were made using Ultimaker PLA. PLA is a biodegradable, cheap and straightforward to use material offering high stiffness. It is the most appropriate material for rapidly prototyping initial development prototypes, as required for the present work. Measurement geometries and their corresponding outer cups were manufactured with XSTRAND^{TM} GF30-PP—glass fibre reinforced polypropylene (Owens Corning, Toledo, OH, USA). This material offers much higher chemical resistance and thermal stability compared to PLA, making it suited for use with chemically aggressive substances, such as ZM in nitric acid. Table 1 compares the mechanical and other properties of these materials.

Table 1. Comparison of the parameters of Ultimaker PLA and XSTRAND^{TM} GF30-PP 3D printing filaments [23,24].

	Ultimaker PLA	XSTRAND^{TM} GF30-PP
Tensile strength at yield (MPa)	49.5	60.0
Flexural strength at yield (MPa)	103.0	83.0
Maximum usable temperature (°C)	50	120
Chemically resistant	No	Yes

3. Device Development and Results

The present section describes the new device, its calibration and evaluation results.

3.1. Description of Designed Device

The device is a rotational viscometer, using bob and vane geometries with outer cups. It uses a spring-loaded mechanism to measure the reaction torque from the analysed substance as a function of the deflection of the drive system.

Figure 2 illustrates the mechanism and the major parts used in it. A stepper motor (RS Components Ltd, Corby, UK, RS PRO 535-0372) drives a measurement geometry (described further below and pictured in Figure 3) with a constant rotational velocity in the substance it is analysing. When the measurement geometry shears the sample, the reaction torque turns the entire drive assembly in the opposite direction, as the motor is held in place by a bearing that allows it to rotate around the same axis as the measurement geometry. A spring connecting the drive assembly to the stationary frame ensures that the mechanism does not spin continuously, i.e., it only deflects proportionally to the shear stress of the material that the measurement geometry shears.

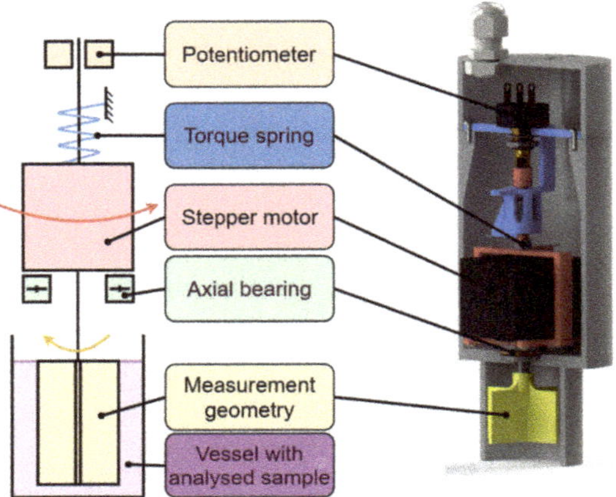

Figure 2. Functional schematic and a render of the prototype.

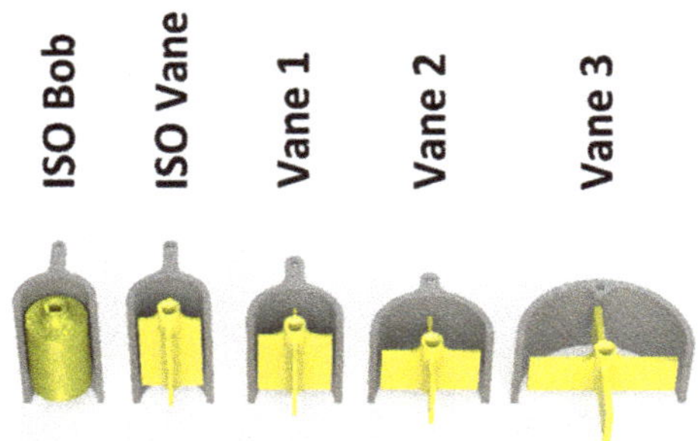

Figure 3. Cutaway render of all the geometries used with the prototype.

Two of the geometries and their respective cups used with the device are designed based on the dimensions proposed by the ISO 3219 standard, widely used in the design of conventional benchtop rheometers [25]. They are manufactured with the same dimensions as the V25 vane and C25 bob used on the Bohlin CVO100 rheometer. The device, however, diverges from the ISO standard in two main regards. One, since it is deployed by inserting it in a sample, the cup is open on the bottom. This allows the material to enter the cup and surround the measurement geometry as the device is inserted into the material. Secondly, the immersion height is not controlled by sampling but by deploying the device. In the experiments described below, the material height is controlled in the test material vessel.

In potential deployment scenarios, the material height may not be sufficient to fully immerse long geometries. For this reason, three other geometries have been designed and used, diverging from the ISO standard. Another consideration is preventing the slip of the geometry when it rotates in the material. Vanes are often used with suspensions and higher viscosity samples, as they slip less than bobs when shearing these samples. Finally, inserting the geometry itself affects the structure of the sample. Vanes penetrate samples much easier than bobs. Based on all of these considerations, all three non-standard geometries (i.e., Vane 1, 2 and 3 in Figure 3) take the form of four-bladed vanes.

As described in the ISO standard, the representative shear rate in the middle of the gap between the geometry and the cup $\dot{\gamma}_{rep}$ is determined as shown in Equation (1).

$$\dot{\gamma}_{rep} = \Omega \cdot \frac{1 + \left(\frac{r_e}{r_i}\right)^2}{\left(\frac{r_e}{r_i}\right)^2 - 1},\tag{1}$$

where Ω is the angular velocity of the geometry, r_e is the radius of the outer cylinder (cup) and r_i is the radius of the inner cylinder (geometry). It is apparent that the shear rate is dependent on the ratio of the dimensions of the geometry and the cup, and thus, the ratio has been maintained constant with all the geometries. To accommodate potential low material heights, the non-standard vanes have been designed to be shorter and thus require less material to be fully immersed. To ensure that the torque is not below the detection threshold when performing the measurement, the radius of the geometry has been increased in comparison with the ISO geometries.

A summary of the geometry dimensions that have been manufactured and used with the prototype is outlined in Table 2, while the geometry and cup configuration is illustrated in Figure 3.

Table 2. Comparison of the dimensions of geometries used with the device.

	ISO Bob	ISO Vane	Vane 1	Vane 2	Vane 3
Cup Radius r_e (mm)	13.75	13.75	16.50	22.00	33.00
Geometry Radius r_i (mm)	12.50	12.50	15.00	20.00	30.00
Geometry Length L (mm)	37.50	37.50	30.00	20.00	15.00
L/r_i (-)	3.00	3.00	2.00	1.00	0.50
Necessary Sample Height (mm)	52.50	52.50	42.50	32.50	27.50

The outer case and internal linkages were 3D printed using a fused deposition modelling printer with Ultimaker PLA. The bob and vane geometries and the outer cups were 3D printed with XSTRAND^{TM} GF30-PP. Overall, the largest configuration (ISO vane or bob) of the device is 219 mm long (i.e., edge of the cable gland to end of the geometry cup) and 75 mm in diameter at the widest point.

The prototype was designed to remove as many sensitive electronic components from the deployment environment. Hence, the electronic circuitry consists of two main assemblies (Figure 4).

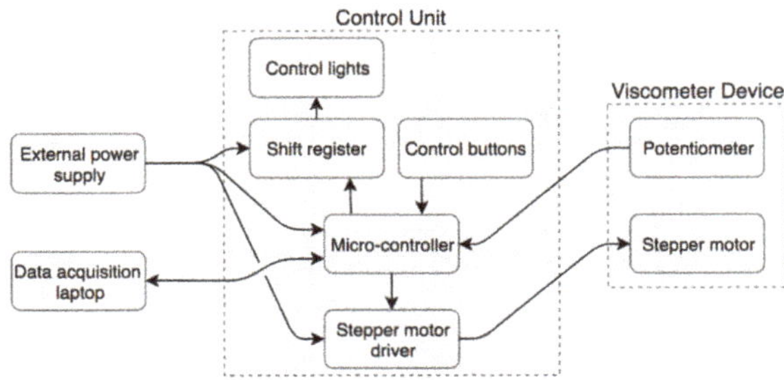

Figure 4. Schematic of the electronic components used with the prototype.

The Control Unit houses all the data acquisition and control electronics necessary to drive the mechanism and read the data from an on-board sensor. All of the components used are COTS products. A Cortex-M4F (180 Mhz rated core speed) based microcontroller was selected as it supports the C++ based programming language, has integrated digital-to-analogue conversion (DAC) functionality and 16-bit resolution analogue-to-digital conversion (ADC) functionality. It controls the stepper motor through a Pololu DRV8834 stepper motor driver (Pololu Corporation, Las Vegas, NV, USA) and reads the output of the Bourns 6630S0D-C28-A103 continuous turn potentiometer (Bourns, Riverside, CA, USA) in the viscometer device. Furthermore, it reads two buttons used to start and stop the pre-tensioning and measurement procedures and controls the 74HC595 shift register used to display the current state of the prototype through 4 LED indicators on the top panel of the Control Unit. All of the Control Unit components are in a PLA housing. The overall dimensions of the housing are 100 mm × 80 mm × 53 mm. All of the units can be seen in Figure 5.

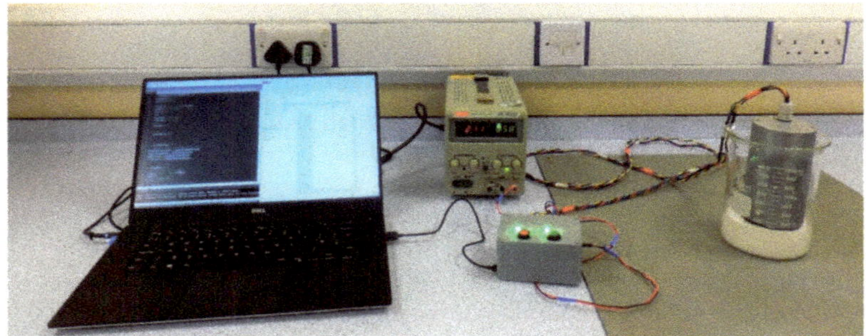

Figure 5. Experimental setup with the prototype. Devices from left to right: laptop, external power supply, control unit, viscometer prototype device in a beaker immersed in a sample of TiO_2.

The prototype is programmed with two procedures. One is to zero the mechanism to establish the origin of the measurement. This procedure runs the geometry in the opposite direction of the measurement rotation in a ramp-up of rotational velocity, which deflects the mechanism. When the mechanism deflects beyond a pre-programmed value, the program switches to a ramp down to allow the mechanism to settle, in what will be used as a zero position. This procedure serves to eliminate all potential deflection and tensioning of the mechanism during deployment.

The second procedure is the measurement itself. The prototype is programmed to supply a ramp-up of rotational velocities in integer values. The stepper motor supplies this velocity for a total of 12 s, measuring the deflection of the mechanism in the last six seconds to allow the mechanism to fully deflect before the measurement. In those six seconds, the microcontroller reads and saves the value from the potentiometer every 400 ms, logs the time since it started the measurement procedure and saves the rotational velocity value it is currently supplying to the geometry.

3.2. Calibration Procedure

A silicone viscosity standard oil (Paragon Scientific VIS-RT1K-600) was used as the calibration material. A baseline measurement was performed using a Bohlin CVO100 rheometer with this oil, using a logarithmically spaced shear rate ramp from 0.1 to 100 s^{-1} over 30 s, with a DIN standard V25 vane a C25 bob and cup configurations. Five runs were made with both geometries, with the averaged shear stress results and the linear fits of data presented in Figure 6. The results are plotted as a function of rotational velocity since the prototype device supplies rotational velocity in revolutions per minute.

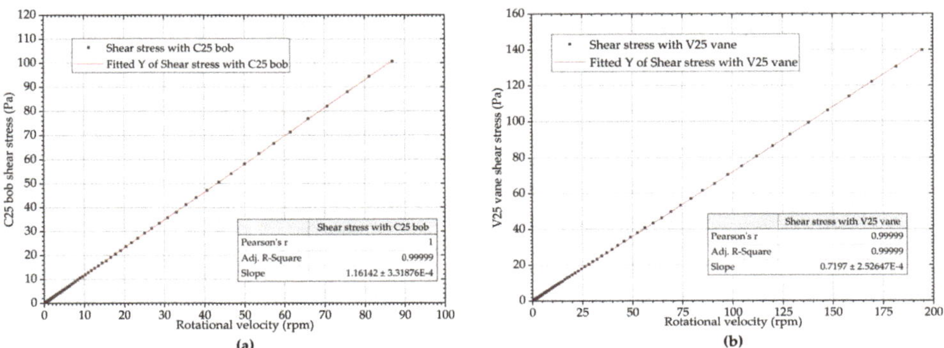

Figure 6. Shear stress as a function of the rotational velocity of silicone viscosity standard oil on a commercial rheometer with: (**a**) C25 bob and cup configuration; (**b**) V25 vane and cup configuration.

The same calibration oil was then used with the prototype and all of its geometries. The prototype supplied rotational velocities in the range of 1 to 100 rpm with the bob and 2 to 200 rpm with the vanes (corresponding to roughly 1 to 100 s^{-1} on the commercial rheometer). The spacing is approximately logarithmic, limited by the fact that the prototype can supply only integer values of revolutions per minute. Five runs with the standard oil were performed for each geometry and the raw data were fitted with linear functions, as illustrated by Figure 7.

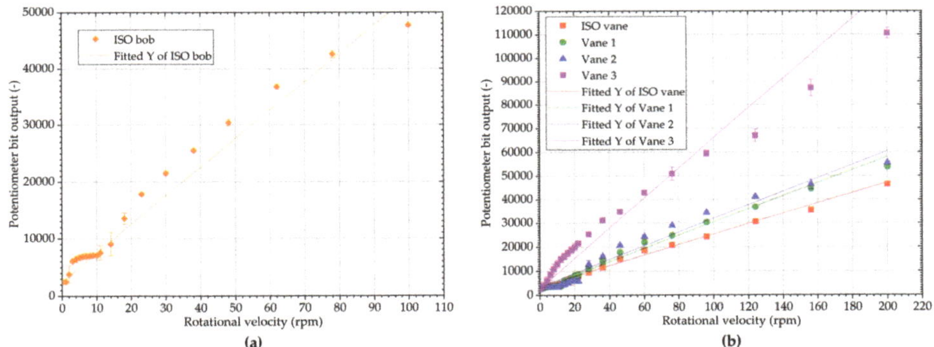

Figure 7. Potentiometer bit output as a function of revolutions per minute of silicone viscosity standard oil with the prototype with (**a**) C25 bob and cup configuration; (**b**) Selection of vanes.

The raw data collected with the potentiometer in the prototype indicates the torque applied to the vane. The raw data were fitted with a linear function and, using data collected with the conventional rheometer, the relationship between the raw data and expected shear stress based on the conventional rheometer measurement was determined. Using this approach, the conversion constants between raw data and shear stress were obtained for each geometry. Summary of calibration data can be seen in Table 3.

Table 3. Summary of calibration data.

	ISO Bob	ISO Vane	Vane 1	Vane 2	Vane 3
R2 of raw data fit (-)	0.9826	0.9970	0.9968	0.9786	0.9793
Slope of prototype raw data fit (-)	502.73	218.72	276.74	285.94	633.86
Slope of commercial rheometer data fit (Pa)	1.1614		0.7197		

Figure 8 summarises the calibration results for all five geometries. Vane 3 offers the best low shear stress detection rate. However, the linearity is limited in the higher shear stress regions. Smaller vanes, namely the ISO and Vane 1 geometries exhibit better repeatability and linearity in higher rotational velocities, but as the torques are lower, low shear stress detection is relatively poor. The bob geometry and Vane 2 exhibit the poorest results overall. With the bob geometry, this can be attributed to the potentially highest sensitivity to axial misalignment of the bob and the cup, with the results being affected by the uneven gap between the bob and the cup.

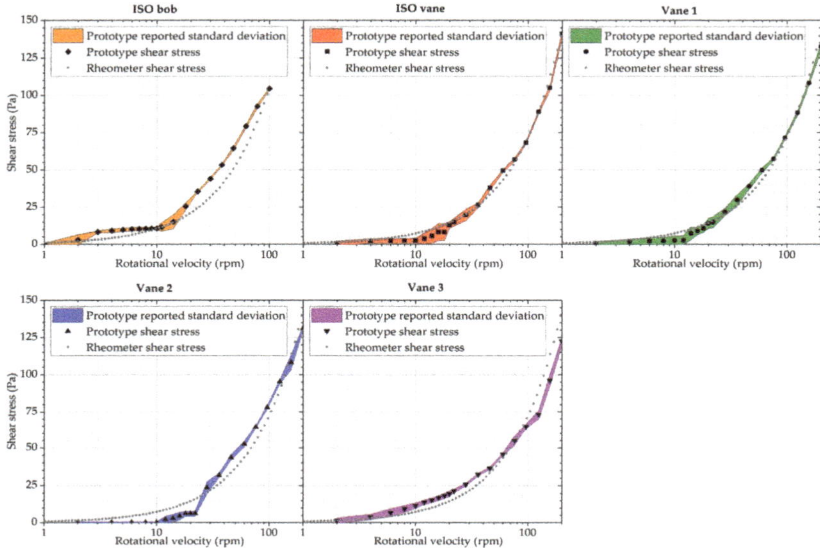

Figure 8. Shear stress results obtained with all five prototype geometries during the calibration procedure, in comparison to a conventional rheometer.

All the geometries exhibit repeatable, systematic deviations from the expected linear response to the silicone oil. Therefore, a compensation procedure was created to remove the nonlinearities. For all the geometries, the relative deviation from the expected value of shear stress was plotted against the shear stress measured with the prototype. Polynomial functions were fitted to the values of the deviation as a function of the measured shear stress, as illustrated in Figure 9. To ensure a good fit using low order functions, only data in the range necessary to be compensated were fitted. These functions are subsequently used to determine the compensation value for the measured results. Actual reported shear stress values are, therefore, defined as the measured shear stress minus the value of the polynomial function at that measured shear stress.

It is important to note that the results obtained with this procedure must be used carefully and that the compensation should only be used within the range of the fitted data, see Figure 9. The results discussed below illustrate instances in which the procedure appears to be appropriate or not.

One aspect of using vanes in rotational rheology that needs to be considered when interpreting the results are secondary flows and vortices appearing in the sample. Research on quantifying the onset of secondary flows using vanes is limited and only applied to one phase materials. Onset of Taylor vortices can be described using the Taylor number. For concentric cylinders, with the outer cylinder stationary and the inner cylinder rotating, in the middle of the gap between the geometries, the Taylor number T_a is often calculated as shown in Equation (2) [26].

$$Ta = \frac{r_i(r_e - r_i)^3 \Omega_i^2}{\nu^2},$$ (2)

where ν is the kinematic viscosity of the sample.

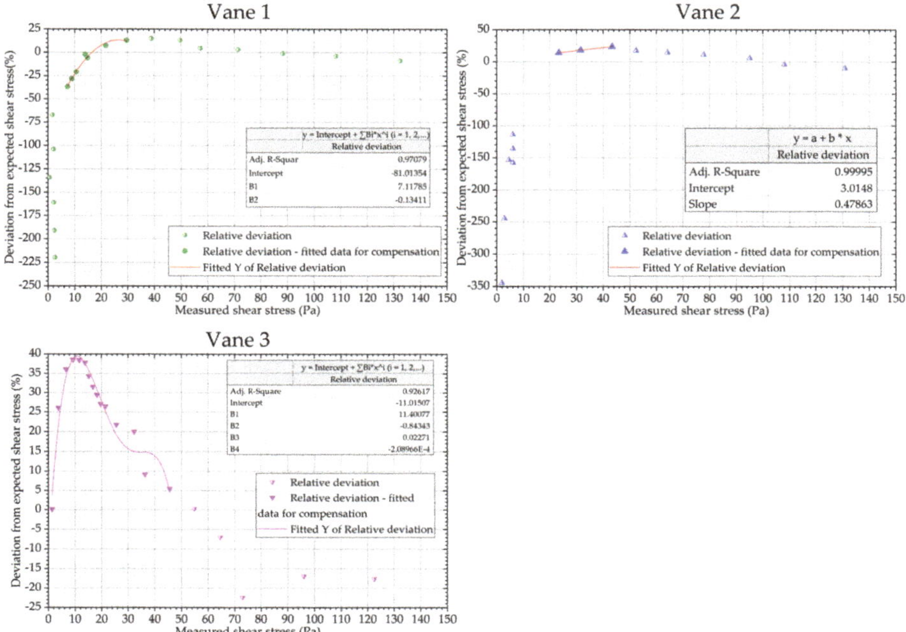

Figure 9. Non-linearity of the prototype with Vanes 1, 2 (first five values are off the scale and not used in calculations) and Vane 3 using a silicone viscosity standard oil and fitting functions for compensation.

The onset of turbulent flow in the same setup as considered in the calculation of the Taylor number is referred to as the rotational Reynolds number and can be calculated as illustrated in Equation (3).

$$Re = \frac{r_i(r_e - r_i)\Omega_i}{\nu}.$$ (3)

It is important to note that these parameters have been validated and used with concentric cylinders and their validity for vanes with particulate suspensions has not been investigated in the literature to date. Rotational rheology using vanes assumes the material shears along the blades of the vanes, and thus vanes are often described using the same principles as concentric cylinders. Current research and rheology laboratory equipment suppliers suggest that vanes are more susceptible to both turbulent flows and Taylor vortices. However, it is apparent from Equations (2) and (3) that it is assumed here that both vanes and bob geometries exhibit the same behaviour. Both Taylor and Reynolds numbers have been calculated using the viscosity and angular velocity determined on the rheometer with the calibration silicone oil and the dimensions of the geometries used with the prototype. The results are illustrated in Figure 10.

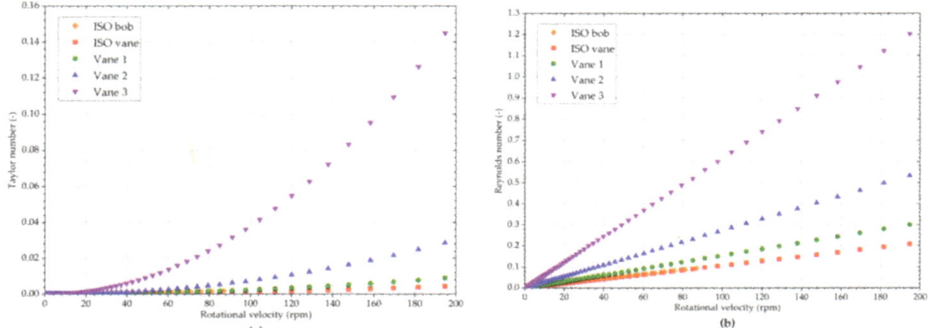

Figure 10. Potential indicators of secondary effects when measuring standard silicone oil: (**a**) Taylor number for all geometries; (**b**) Reynolds number for all geometries.

3.3. Measurements with TiO₂ Suspension

The titanium dioxide suspension is low viscosity, and the shear stress is below the detection level of most geometries, with the exception of Vane 3. Using this geometry, as shown by Figure 11, the uncompensated data aligns with the expected values. However, at approximately 80 rpm, the slope of the curve changes and for high shear rate regions the prototype starts to report values higher than expected. The most likely explanation is secondary flows starting to appear in the analysed material. It is apparent that the compensation procedure is not appropriate for low viscosity sample such as used here, as it causes under-reporting.

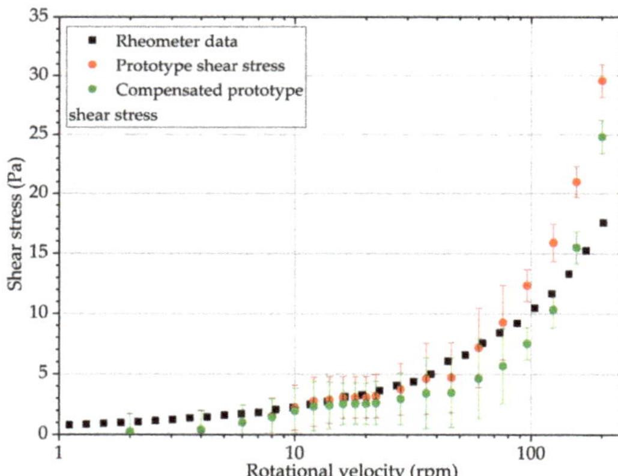

Figure 11. Shear stress results obtained with a rheometer, the prototype (using vane 3) and compensated prototype values with titanium dioxide suspension.

The Taylor and Reynolds numbers calculated using the rheometer collected data, but with the dimensions of Vane 3, are illustrated by Figure 12. Secondary flow effects appear to start at around 80 rpm, corresponding to *Ta* = 6 and *Re* = 7.5. These values are low, compared to reported thresholds indicating the onset of secondary effects in current literature.

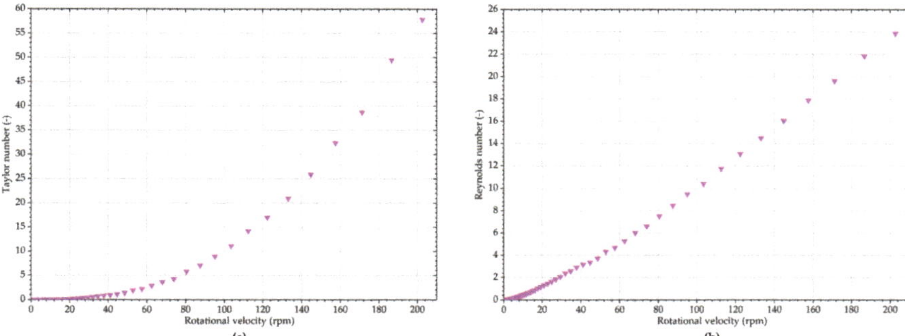

Figure 12. Potential indicators of secondary effects when measuring titanium dioxide suspension with Vane 3: (**a**) Taylor number; (**b**) rotational Reynolds number.

3.4. ZM Suspension

ZM suspension has a much higher viscosity in the low rotational velocity regions. It is also demonstrably shear thinning. It is apparent from Figure 13 that the values obtained with Vane 2 and Vane 3 are higher than expected. With Vane 2, this can be attributed to the relatively poor calibration results. With Vane 3, some over-reporting is expected due to the non-linearity of the response of the prototype. Vane 1 appears to offer the best approximation of the results without compensating the results. All three geometries report shear thinning behaviour.

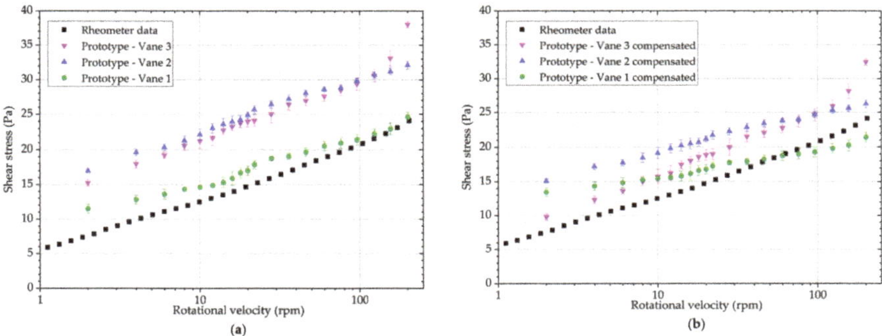

Figure 13. Shear stress results obtained with the rheometer and the prototype with Vanes 1, 2 and 3: (**a**) Measured values; (**b**) Compensated values.

The compensation procedure does not seem appropriate for Vane 1 in this case, as it changes the slope of the curve. Vane 2 is affected similarly. However, the values obtained with the prototype are still higher than expected across the measurement range. The procedure works well for Vane 3, as the slope of the data is still the same, but results are much closer to the expected values.

The potential onset of secondary flow seems to be delayed as opposed to the lower viscosity sample. This is consistent with the calculations of the Taylor and Reynolds numbers, as increased viscosity is expected to delay the onset of secondary flows. The secondary flow effects appear only with Vane 3, from approximately 150 rpm, corresponding to $Ta = 3.5$ and $Re = 6$, as illustrated in Figure 14. These values are similar to those obtained with titanium dioxide suspension.

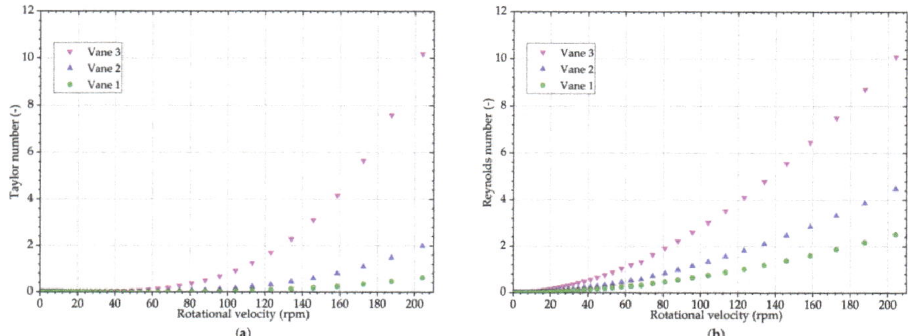

Figure 14. Potential indicators of secondary effects when measuring zirconium molybdate suspension with Vane 1, 2 and 3: (**a**) Taylor number; (**b**) rotational Reynolds number.

4. Discussion and Conclusions

This paper has described the development of a relatively compact and robust device to analyse the rheological properties of sludge without the need to sample materials. Analysis of a sludge test material has been performed to create a suitable benchmark material for the rheological measurements with the prototype.

The results presented in this paper show the advantages and disadvantages of the proposed mechanism and measurement geometries to obtain shear behaviour data of suspensions. The prototype device and all necessary electronics are all either COTS components, or 3D printed, and the total price of the prototype is less than £250. The device also only requires freeware programs to run and transfer data to a PC.

The calibration procedure shows the non-linearity of all geometries. Non-standard geometries offer responses that can be compared with ISO/DIN standard geometries. Larger geometries offer beneficial low shear stress responses, making them suitable for low viscosity samples and potentially for yield stress analysis. The prototype is able to reproduce measurements of lower viscosity suspension samples using the calibration and compensation methods proposed in this article.

No secondary flow effects were observed with any of the geometries using the 1Pa·s silicone standard oil. Potential secondary flow effects were observed in the suspensions at angular velocities over approximately 100 rpm. However, as the tool is being developed for ascertaining low shear rate behaviour, these effects are not prohibitive. Further investigation would be necessary to quantify the onset of these effects. The present results indicate that secondary flows might start appearing at very low Taylor and rotational Reynolds numbers <10.

Vane 3 offers the most comparable results, with a compensation procedure appropriate for the zirconium molybdate sample. Shear-thinning behaviour was correctly reproduced with all three vanes on the ZM sample, and the compensation procedure lowers the deviation from the expected results considerably. A more advanced mathematical model could further improve the compensation procedure. The performance indicates that the prototype would be suitable for deployment, especially considering the low manufacturing costs and the potential to reuse components. Vane 3 is the obvious candidate as it requires the least material to be present in the deployment area and provides the most accurate and consistent results. However, it also requires careful interpretation of high shear rate results due to the possibility of the onset of secondary flows.

The device is compact enough to fit through a 75 mm opening, maximising deployment capabilities. The deployment of the device can be accomplished in various ways, depending on the deployment area. The first option is manual deployment, which would be suitable for smaller containers that can be approached by operators. The device would simply be lowered into the container using an extension rod. The second option relates to the assessment of settled bed behaviour in larger tanks.

Multiple underwater remotely operated vehicles (ROVs) have been designed and deployed on various sites and are capable of carrying payloads such as the proposed device [27]. The prototype could be carried and detached from the carrying platform when it reaches the required position. Lastly, for remote tanks, pipe crawler and tracked robots have been used to carry sensors and other payloads and these platforms could deploy the device in harder to access areas [28,29]. In all instances, the tether would serve not only a data transferring function but also as a means of recovery of the device from these areas. The device is light (0.5 kg in the laboratory scale configuration), and the casing is robust enough to support a recovery procedure using a tether. The simple assembly would make it easy to reuse the internal components if necessary and decontaminate the outer casing, cups and geometries after use. If the device cannot be decontaminated, the low price supports the case to use the device only once before disposal.

Potential future development of the device could include lowering friction in the mechanism, which would improve low shear stress detection and repeatability of the measurements. A more complex mathematical model for the non-linearity compensation would be beneficial for the interpretation of the results. Lastly, understanding the onset of secondary flows in various particulate suspensions would be necessary to increase the operating range of the instrument.

Overall, the precision and performance of the proposed mechanism makes it suitable for preliminary analysis of sludge in situ. The capability of the current stage of the prototype would make it useful for determining the rough properties of the materials, to inform further decisions as necessary in the upcoming steps of Post Operational Clean Out (POCO).

Author Contributions: T.F. designed and developed the device. T.F. developed the methodology and interpreted the results with the assistance of J.M.D., D.C. and S.D.M. T.F. wrote the paper with assistance of J.M.D., D.C., S.D.M. and C.J.T.

Funding: This research was jointly funded by the CINDe consortium and by the UK Engineering and Physical Sciences Research Council (EPSRC), via grant EP/R02572X/1.

Acknowledgments: This work has been supported by the Centre for Innovative Nuclear Decommissioning (CINDe), which is led by the National Nuclear Laboratory, in partnership with Sellafield Ltd. and a network of Universities that includes the University of Manchester, Lancaster University, the University of Liverpool and the University of Cumbria.

Conflicts of Interest: The authors declare no conflict of interest.

References

1. NIRAB. The UK Civil Nuclear R&D Landscape Survey (February 2017). Available online: http://www.nirab. org.uk/media/10671/nirab-123-4.pdf (accessed on 19 March 2019).
2. GOV.UK End of Reprocessing at Thorp Signals New Era for Sellafield. Available online: https://www.gov.uk/ government/news/end-of-reprocessing-at-thorp-signals-new-era-for-sellafield (accessed on 19 March 2019).
3. Sellafield Ltd. Corporate Plan 2016/17. Available online: https://www.gov.uk/government/uploads/system/ uploads/attachment_data/file/627566/SEL11098_corporate-plan_web.pdf (accessed on 19 March 2019).
4. OECD. *R&D and Innovation Needs for Decommissioning Nuclear Facilities*; Radioactive Waste Management, OECD Publishing: Paris, France, 2014; ISBN 9789264222199.
5. Keele, B.D.; Addleman, R.S.; Blewett, G.R.; Mcclellan, C.S.; Troyer, G.L. Non-destructive in situ measurement of radiological distributions in Hanford Site waste tanks. *IEEE Trans. Nucl. Sci.* **1996**, *43*, 1821–1826. [CrossRef]
6. Maley, S.; Jones, H.; Whittaker, W. Novel Radiometry for In-Pipe Robotic Inspection of Holdup Deposits in Gaseous Diffusion Piping. In Proceedings of the WM2019 Conference, Phoenix, AZ, USA, 3–7 March 2019.
7. Bird, B.; Griffiths, A.; Martin, H.; Codres, E.; Jones, J.; Stancu, A.; Lennox, B.; Watson, S.; Poteau, X. Radiological Monitoring of Nuclear Facilities: Using the Continuous Autonomous Radiation Monitoring Assistance Robot. *IEEE Robot. Autom. Mag.* **2018**. [CrossRef]
8. Tsitsimpelis, I.; Taylor, C.J.; Lennox, B.; Joyce, M.J. A review of ground-based robotic systems for the characterization of nuclear environments. *Prog. Nucl. Energy* **2019**, *111*, 109–124. [CrossRef]

9.	Baudelet, M. 17 Applications of laser spectroscopy in nuclear research and industry. In *Laser Spectroscopy for Sensing*; Woodhead Publishing: Sawston, UK, 2014; pp. 522–543.

10.	Martin, M.Z.; Allman, S.; Brice, D.J.; Martin, R.C.; Andre, N.O. Exploring laser-induced breakdown spectroscopy for nuclear materials analysis and in-situ applications. *Spectrochim. Acta Part B At. Spectrosc.* **2012**, *74*, 177–183. [CrossRef]

11.	Griffiths, A.; Dikarev, A.; Green, P.R.; Lennox, B.; Poteau, X.; Watson, S. AVEXIS-Aqua vehicle explorer for in-situ sensing. *IEEE Robot. Autom. Lett.* **2016**, *1*, 282–287. [CrossRef]

12.	Mould, S.; Barbas, J.; Machado, A.V.; Nóbrega, J.M.; Covas, J.A. Measuring the rheological properties of polymer melts with on-line rotational rheometry. *Polym. Test.* **2011**, *30*, 602–610. [CrossRef]

13.	Klein, C.O.; Spiess, H.W.; Calin, A.; Balan, C.; Wilhelm, M. Separation of the nonlinear oscillatory response into a superposition of linear, strain hardening, strain softening, and wall slip response. *Macromolecules* **2007**, *40*, 4250–4259. [CrossRef]

14.	Poulos, A.; Renou, F.; Jacob, A.; Koumakis, N.; Petekidis, G. Large amplitude oscillatory shear (LAOS) in model colloidal suspensions and glasses: Frequency dependence. *Rheol. Acta* **2015**, *54*, 715–724. [CrossRef]

15.	Schermeyer, M.; Sigloch, H.; Bauer, K.C.; Oelschlaeger, C.; Hubbuch, J. Squeeze flow rheometry as a novel tool for the characterization of highly concentrated protein solutions. *Biotechnol. Bioeng.* **2016**, *113*, 576–587. [CrossRef] [PubMed]

16.	Kádár, R.; Naue, I.F.C.; Wilhelm, M. First normal stress difference and in-situ spectral dynamics in a high sensitivity extrusion die for capillary rheometry via the 'hole effect'. *Polymer* **2016**, *104*, 193–203. [CrossRef]

17.	Morita, A.T.; Toma, M.S.; Paoli, M.-A. De Low cost capillary rheometer, transfer molding and die-drawing module. *Polym. Test.* **2006**, *25*, 197–202. [CrossRef]

18.	Paul, N.; Biggs, S.; Edmondson, M.; Hunter, T.N.; Hammond, R.B. Characterising highly active nuclear waste simulants. *Chem. Eng. Res. Des.* **2013**, *91*, 742–751. [CrossRef]

19.	Pease, L.F.; Daniel, R.C.; Burns, C.A. Slurry rheology of Hanford sludge. *Chem. Eng. Sci.* **2019**, *199*, 628–634. [CrossRef]

20.	Paul, N. *Characterisation of Highly Active Nuclear Waste Simulants*; University of Leeds: Leeds, UK, 2014.

21.	Botha, J.A.; Ding, W.; Hunter, T.N.; Biggs, S.; Mackay, G.A.; Cowley, R.; Woodbury, S.E.; Harbottle, D. Quartz crystal microbalance as a device to measure the yield stress of colloidal suspensions. *Colloids Surf. A Physicochem. Eng. Asp.* **2018**, *546*, 179–185. [CrossRef]

22.	Paul, N.; Hammond, R.B.; Hunter, T.N.; Edmondson, M.; Maxwell, L.; Biggs, S. Synthesis of nuclear waste simulants by reaction precipitation: Formation of caesium phosphomolybdate, zirconium molybdate and morphology modification with citratomolybdate complex. *Polyhedron* **2015**, *89*, 129–141. [CrossRef]

23.	PLA|Professional 3D Printing Made Accessible|Ultimaker. Available online: https://ultimaker.com/en/resources/49911-pla#technical (accessed on 17 April 2019).

24.	XSTRAND™ GF30-PP-Owens Corning Composites. Available online: https://www.owenscorning.com/composites/product/xstrand-gf30-pp (accessed on 17 April 2019).

25.	British Standards Institution. Plastics—Polymers/Resins in the Liquid State or as Emulsions or Dispersions—Determination of Viscosity Using a Rotational Viscometer with Defined Shear Rate. EU Patent EN ISO 3219:1995, 15 March 1994.

26.	White, F.M. *Fluid Mechanics*, 7th ed.; McGraw-Hill: New York, NY, USA, 2011; ISBN 978-0-07-352934-9.

27.	Nancekievill, M.; Jones, A.R.; Joyce, M.J.; Lennox, B.; Watson, S.; Katakura, J.; Okumura, K.; Kamada, S.; Katoh, M.; Nishimura, K. Development of a Radiological Characterization Submersible ROV for Use at Fukushima Daiichi. *IEEE Trans. Nucl. Sci.* **2018**, *65*, 2565–2572. [CrossRef]

28.	Tibrea, S.; Nance, T.; Kriikk, E. Robotics in hazardous environments—Real deployments by the savannah river national laboratory. *J. South Carol. Acad. Sci.* **2011**, *9*, 5.

29.	Rood, M.; Kelly, D. Hanford Double Shell Tank Integrity Inspection–Integration of Customized Robotics. In Proceedings of the Waste Management Symposia, Phoenix, AZ, USA, 18–22 March 2018.

Article

Low Voltage High-Energy α-Particle Detectors by GaN-on-GaN Schottky Diodes with Record-High Charge Collection Efficiency

Abhinay Sandupatla [1]*, Subramaniam Arulkumaran [2,3], Kumud Ranjan [2], Geok Ing Ng [1,2]*, Peter P. Murmu [4], John Kennedy [4], Shugo Nitta, Yoshio Honda [4], Manato Deki [3] and Hiroshi Amano [3]

[1] School of Electrical and Electronics Engineering, Nanyang Technological University, Singapore 639798, Singapore
[2] Temasek Laboratories @ NTU, Research Techno Plaza, 50 Nanyang Drive, Singapore 639798, Singapore; Subramaniam@ntu.edu.sg (S.A); KRanjan@ntu.edu.sg (K.R.)
[3] Center for Integrated Research of Future Electronics (CIRFE), IMaSS, Nagoya University, Nagoya 464-8603, Japan; nitta@nagoya-u.jp (S.N); honda@nuee.nagoya-u.ac.jp (Y.H.); deki@nuee.nagoya-u.ac.jp (M.D.); amano@nuee.nagoya-u.ac.jp (H.A.)
[4] National Isotope Center, GNS Science, Lower Hutt 5010, New Zealand; P.Murmu@gns.cri.nz (P.P.M.); J.Kennedy@gns.cri.nz (J.K.)
* Correspondence: abhinay001@e.ntu.edu.sg (A.S.); eging@ntu.edu.sg (G.I.N.)

Received: 10 October 2019; Accepted: 19 November 2019; Published: 21 November 2019

Abstract: A low voltage (−20 V) operating high-energy (5.48 MeV) α-particle detector with a high charge collection efficiency (CCE) of approximately 65% was observed from the compensated $(7.7 \times 10^{14}\,/\text{cm}^3)$ metalorganic vapor phase epitaxy (MOVPE) grown 15 µm thick drift layer gallium nitride (GaN) Schottky diodes on free-standing n+-GaN substrate. The observed CCE was 30% higher than the bulk GaN (400 µm)-based Schottky barrier diodes (SBD) at −20 V. This is the first report of α−particle detection at 5.48 MeV with a high CCE at −20 V operation. In addition, the detectors also exhibited a three-times smaller variation in CCE (0.12 %/V) with a change in bias conditions from −120 V to −20 V. The dramatic reduction in CCE variation with voltage and improved CCE was a result of the reduced charge carrier density (CCD) due to the compensation by Mg in the grown drift layer (DL), which resulted in the increased depletion width (DW) of the fabricated GaN SBDs. The SBDs also reached a CCE of approximately 96.7% at −300 V.

Keywords: high-energy α-particle detection; low voltage; thick depletion width detectors

1. Introduction

The measurement of the energy spectrum of charged particles plays an important role in studying the fusion reactions of nuclear reactors and in particle physics research conducted at facilities like the Large Hadron Collider. Gallium Arsenide (GaAs) has been the primary contender as an alternative to Si-based detectors. However, due to its low displacement energy (E_d) of 10 eV [1], the lifetime of the GaAs detector is limited. Low E_d results in the easier displacement of Ga and As atoms from their respective crystal lattice, thereby weakening their lattice structure. Gallium Nitride (GaN) was considered as an alternative in high-energy charged particle detection due to its higher E_d (20 eV) [1]. The high E_d of GaN would improve the radiation tolerance of the GaN detector, as reported by B.D. Weaver et al., by comparing the radiation damage on GaN and GaAs [2]. The report also determined that GaN could withstand twice the dosage in comparison to GaAs. GaN also has a larger bandgap of 3.4 eV [3] in comparison to GaAs (1.42 eV) and Si (1.12 eV), which enables GaN detectors to operate at higher temperatures. The superior material characteristics of GaN have encouraged multiple research groups in exploring applications of GaN devices in radiation detection. GaN devices have performed exceedingly well as α-particle detectors [4–13] and neutron detectors [14–17]. The structure of these GaN detectors can be classified into three types, namely double Schottky contact (DSC) structure, mesa structure and sandwich structure.

The first ever GaN alpha particle detector was realized by Vaitkus et al. [4] with the DSC structure. The detector had a 2 µm GaN epi-layer, which resulted in the detection of 410 keV alpha particles with a 92% charge collection efficiency (CCE). Similarly, other research groups have successfully implemented different types of mesa structures and achieved the higher CCE of 100% due to a lower trapping impurity concentration. Compared to DSC structures and mesa structures, sandwich structures have the potential to generate the thickest depletion widths (DWs), which enables the detection of higher energies. The development of the sandwich structures of GaN detectors was primarily limited by the unavailability of free-standing substrates. With the improvement in the GaN growth technologies, researchers have been able to produce high-quality free-standing GaN substrates and epitaxial films. This has led many research groups to explore the sandwich structure for GaN-based radiation sensors [4–6,8]. GaN detectors with thin epitaxial films detect only fractions of the typical energies emitted by actinides (4 MeV to 6 MeV), such as U-235 (4.268 MeV), ^{241}Am (5.48 MeV) and Pu (4.67 MeV). While different thicknesses of epitaxial drift layers (DLs) (2 µm to 12 µm) were tested, they could only detect energies in the range of 0.5 MeV [4] to 4.5 MeV [5], which is on the lower end of detection requirements. To detect higher energies, researchers increased the DWs of the detector by fabricating them on bulk GaN substrates. These detectors generate a 27 µm DW at very high voltages (-550 V) to detect high energies (5.48 MeV) [6]. The high voltage required to generate a thick DW in bulk GaN-based detectors increases both the complexity and size of the detector, thus severely affecting its portability. In this work, for the first time we design and fabricate GaN-on-GaN Schottky barrier diodes with compensated metalorganic vapor phase epitaxy (MOVPE) grown GaN DL to detect 5.48 MeV α-particles (^{241}Am source), even at -20 V.

2. Design of α-Particle Detector

To design an α-particle detector that works in low-bias conditions and detects high α-particle energies, we employed GaN-on-GaN Schottky barrier diodes (SBDs) with a sandwich structure (see Figure 1). The thin epitaxial GaN layer was very lowly doped, resulting in a thick DW, while the highly doped substrate gave mechanical strength to the detector and formed a good ohmic contact, which helped to reduce the biasing voltage needed.

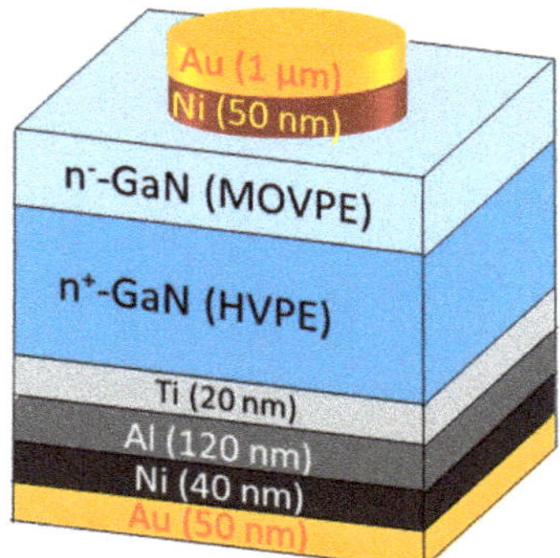

Figure 1. Gallium nitride-on-gallium nitride (GaN-on-GaN) Schottky barrier diodes (SBD) with a sandwich structure for alpha particle detection.

The thickness of the DL plays an important role in determining the maximum thickness of DW, thereby the maximum energy that can be detected. The DL required to detect 5.48 MeV emitted by ^{241}Am was calculated through Stopping and Range of Ions in Matter (SRIM) simulations by considering GaN density (6.1 g/cm^3) and α-particle mass (4.003 amu). Figure 2 shows the SRIM-calculated range of the thickness required in GaN to absorb α-particles with energies between 10 keV and 6 MeV. From this graph, we found that we needed a 14.54 µm DW to detect α-particles with 5.48 MeV. Hence, SBDs with a DL of minimum 15 µm were required for 5.48 MeV α-particle detection.

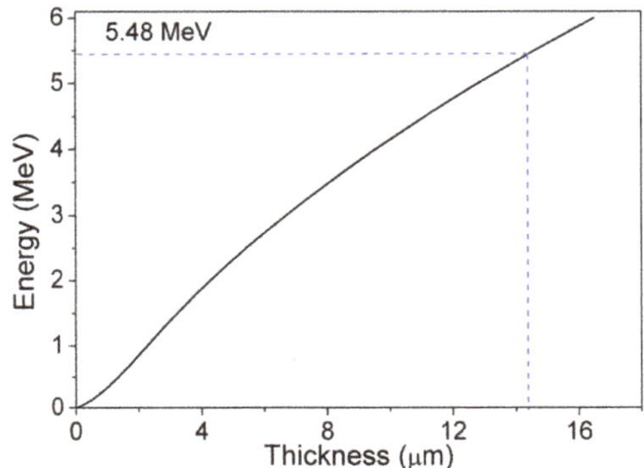

Figure 2. α-particle (^{241}Am, 5.48 MeV) range in GaN calculated by Stopping and Range of Ions in Matter (SRIM).

The 15 µm thick GaN DL was grown by MOVPE at 1080 °C (100 kPa) on hydride vapor phase epitaxy (HVPE)-grown free-standing GaN substrates with a charge carrier density (CCD) of 1×10^{18}/cm^3. The growth rate of 3.5 µm/h was maintained during the growth. The grown DL exhibited low CCD, which was measured using Van der Pauw, Hall and secondary ion mass spectrometry (SIMS) analyses (See Figure 3 and Table 1). From SIMS analysis, the concentrations of different elements present were extracted. While the measured values of O were below the detection limit (3×10^{15} /cm^3), all C, Si, Mg and Fe were present. For the calculation of CCD, we list the extracted concentrations of Si and Mg in Table 1. Though the calculated CCD from SIMS was nominally negligible, the calculated values from SIMS were in a similar range of CCD measured by Hall measurements. Panchromatic cathodoluminescence (CL) measurements were also performed to count the threading dislocation density (TDD), which was an average of 2.65×10^6/cm^2 on the MOVPE-grown GaN DL.

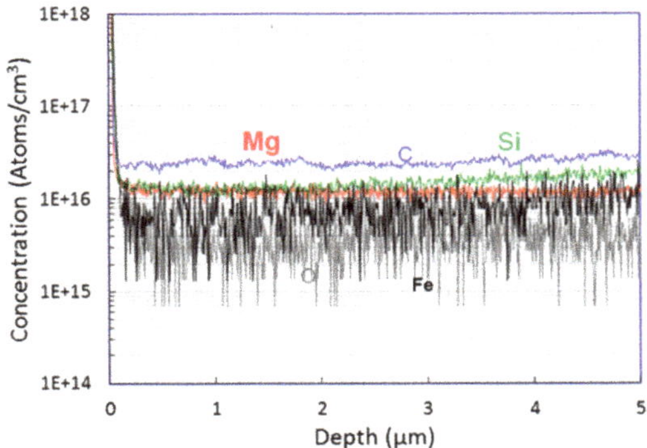

Figure 3. Elemental concentration extracted through secondary ion mass spectrometry (SIMS) at different depths of the drift layer (DL).

Table 1. Concentration of Si and Mg measured through SIMS with the corresponding CCD extracted from SIMS and Hall measurements.

Si (N_D) ($/cm^3$)	Mg (N_A) ($/cm^3$)	CCD = $N_D - N_A$ ($/cm^3$)	
		SIMS	Hall
154.79×10^{14}	147.09×10^{14}	7.7×10^{14}	7.5×10^{14}

3. Detector Fabrication and Measurement Setup

3.1. Detector Fabrication

The fabrication of SBDs started with a thorough cleaning of the GaN-on-GaN wafer with piranha solution and organic cleaning (acetone and isopropanol) and dipping in buffered oxide etchant (BOE) for two minutes to ensure the formation of an excellent metal-semiconductor interface [18]. After the surface preparation, the ohmic contact was formed by depositing Ti/Al/Ni/Au (20/120/40/50 nm) on the N-face (backside) of the wafer, followed by rapid thermal annealing at 775 °C for 30 s in N_2 ambience. The selection of Ti was to form a low-resistance contact as Ti helps generate large amounts of N-vacancies after annealing [18], which increases CCD and promotes tunneling. The second element Al was used to absorb excessive Ti material [19], while Ni was used as a barrier metal, which confines the downward diffusion of the fourth layer (Au) [20]. The top layer of Au was required to protect layers below from oxidization [21]. Multiple SBDs of different sizes were then fabricated by depositing Ni/Au (50/1000 nm) on the Ga-face of the wafer. Ni was selected as the first layer due to the difference in work functions of Ni (5.04 eV) and GaN (4.2 eV) [22], which helps form Schottky contact. After the formation of SBDs, electrical characterization was performed, followed by the dicing of the wafer into individual SBDs. Each SBD was then mounted on a dual in-line package (DIP) by connecting the ohmic contact of the SBD and the ground of the DIP with conductive Ag paste. The Schottky contact was wire bonded using 20 µm diameter Au wires (see Figure 4).

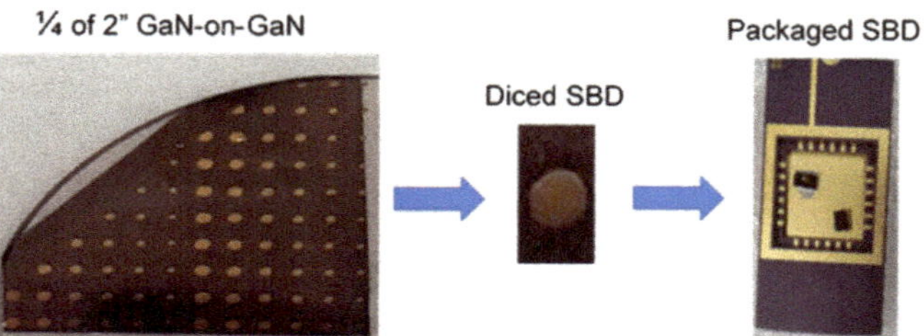

Figure 4. Three stages of sample preparation.

3.2. α-Particle Measurement Setup

^{241}Am was used as a source for the generation of 5.48 MeV α-particles with an active area of 7 mm^2 placed at 8 mm from the detector (as shown in Figure 5). The α-particle source had radionuclides deposited onto a stainless-steel disc of 16 mm diameter, which was held in place by a plastic holder. The GaN SBD detector was connected to a pre-amplifier, amplifier and signal-processing circuit to detect the change in the current flowing through the SBDs due to the interaction with an α-particle. A Si surface barrier detector from ORTEC was used as a reference, along with an ORTEC-671 amplifier for energy calibration.

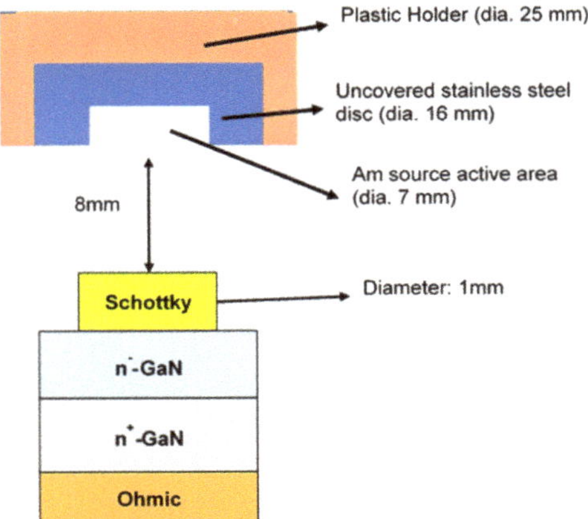

Figure 5. Schematic drawing of Source-Detector measurement setup (not to scale).

4. Results and Discussion

4.1. Current–Voltage (I–V) Characteristics

The *I–V* characteristics were measured using a B1505A power device analyzer at room temperature. Mg-compensation of the GaN DL helped to reduce the CCD by two orders of magnitude, from $4.6 \times 10^{16}/cm^3$ to $7.7 \times 10^{14}/cm^3$. The huge reduction in CCD resulted in the increase of the breakdown voltage by approximately three times, from 462 V to 1480 V, and the reduction of the reverse leakage current by approximately three orders (see Figure 6). A very low reverse leakage current of 3 pA at −20 V bias is low enough for event-by-event counting to acquire α-particle energy spectra. The SBD also exhibited an average ideality factor and Schottky barrier height of 1.03 and 0.79 eV, respectively [23–25]. The near-unity ideality factor signified an excellent metal-semiconductor interface at the Schottky-semiconductor contact [26]. Similarly, the extracted barrier height of 0.79 eV was similar to other reported Ni-based Schottky contacts [27,28].

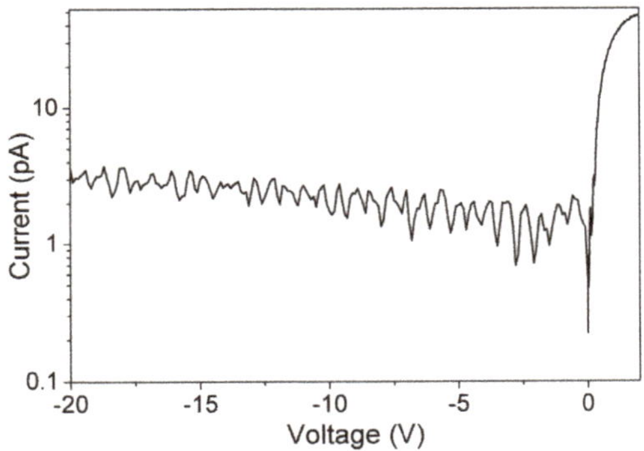

Figure 6. Room temperature *I–V* characteristics of 1 mm diameter GaN SBDs with 15 µm thick compensated DL.

4.2. Capacitance–Voltage (C–V) Characteristics

C–V measurements were also performed to extract the DW of the GaN-on-GaN SBDs. No significant variation in capacitance value was observed for a voltage range of −20 V to 5 V (see Figure 7), which signifies the complete depletion of the DL [10,29].

DW can be extracted from the C–V characteristics using Equation (1):

$$C = \varepsilon_0\varepsilon_r(A/DW)$$ (1)

A uniform DW of approximately 15 µm was measured at all voltages (−20 V to 5 V), which implies the complete DL is depleted even at 0 V.

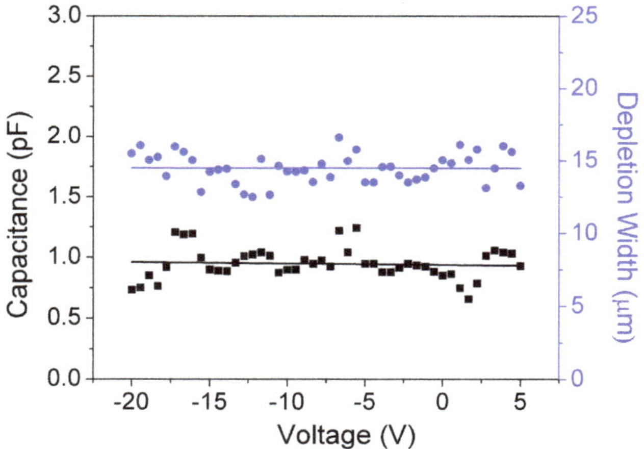

Figure 7. Variation of capacitance and depletion width (DW) with voltage of 0.5 mm diameter GaN SBDs with 15 µm thick Mg-compensated DL.

4.3. Detection of α-Particle Spectra

The performance of an α-particle detector is primarily defined by its CCE. CCE is the ratio of detected energy and incident energy, which is dependent on the DW of the detector. The acquired data was calibrated using a standard Si detector as a reference. In this process, the fabricated GaN detector and the reference Si detector were connected to the same measurement setup separately, without changing the settings. The final detected energy is described by the following equation [6]:

$$E = E_0 + W_{GaN}/W_{Si} \times k \times \text{Channel}$$ (2)

where E is the absorbed energy; E_0 represents energy loss at the metal interface, which was estimated based on the Transport of Ions in Matter (TRIM) simulation to be 183 keV for an Au/Ni (1000/50 nm) contact; k is a calibration factor of the reference Si detector; W_{GaN} (8.9 eV) [30] and W_{Si} (3.6 eV) are the energies required to generate an electron-hole pair in GaN and Si, respectively.

4.3.1. Variation in α-Particle Spectra-Air vs. Vacuum

Figure 8 shows the comparison of the energy spectrum of GaN detectors biased at −100 V measured in a vacuum and in air. About a 7% reduction in CCE was observed when the detectors were measured in air. When an α-particle passes through the vacuum, all its energy is transferred to the detector, resulting in high-energy detection. When α-particles traverse through air, scattering results in loss of energy. This loss in α-particle energy results in a lower CCE. For practical applications, portability and cost are important factors and the need for a vacuum restricts portability and also increases the cost of the system. While the presence of air results in a 7% drop in CCE, compensated SBDs could still be considered for practical applications.

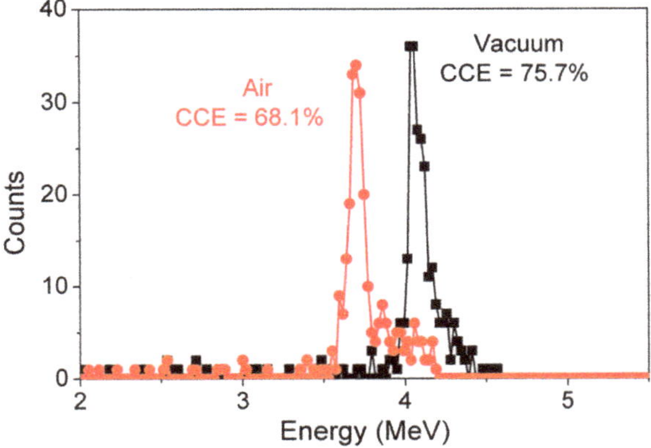

Figure 8. Acquired α-particle energy spectra of GaN SBDs at −100 V under air and in a vacuum.

4.3.2. Low Voltage α-Particle Detection

The α-particle energy spectra obtained under low-bias conditions (−20 to −80V) are shown in Figure 9a. It can be observed that with the decrease in applied bias, the detected energy also decreases. The decrease in detected energy is due to the reduction in DW at lower bias conditions. Reduced DW decreases the path length of the α-particle inside DW, thereby limiting the ability to detect high-energy α-particles. To check detector performance uniformity, four detectors were tested under similar conditions and only a small variation of approximately 2% was observed in measured CCE. Figure 9b shows the variation of CCE with the voltage of our detectors compared with other published bulk GaN detectors (sandwich structures). Our detectors exhibited very low variation in CCE (7%) with a change in voltage (−20 V to −80 V) when compared to the variation shown by a 450 μm thick bulk GaN detector (32.7%) [6] and a 500 μm thick bulk GaN detector (58.1%) [7]. The low variation in CCE is mainly due to the presence of a thick DW even at low-bias conditions, which was achieved by reducing CCD by the compensation of DL.

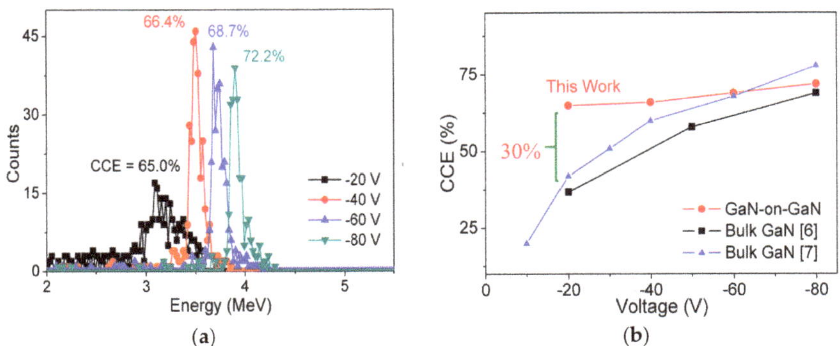

Figure 9. (**a**) Acquired α-particle spectra of GaN SBDs for different applied voltages (−20 V to −80 V) and (**b**) Comparison of measured charge collection efficiency (CCE) of SBDs vs. applied voltages (−20 V to −80 V) with state-of-the-art reported values.

4.3.3. High Voltage α-Particle Detection

Our detectors were also biased at higher voltages to obtain a thicker DW leading to a higher CCE, similar in the range reported by others using bulk GaN detectors [6,7]. Our detector performance improved from 72% at −80 V to 96.7% at −300 V (see Figure 10a), which is the lowest reported voltage at which 5.48 MeV α-particle was successfully detected. The high-voltage performance of these detectors was compared with other published α-particle detectors. Of the many published reports, only Q. Xu et al. reported a high CCE of 100% while detecting a 5.48 MeV α-particle. However, their detectors need to be biased up to −550 V, which is 250 V higher than our detectors (see Figure 10b).

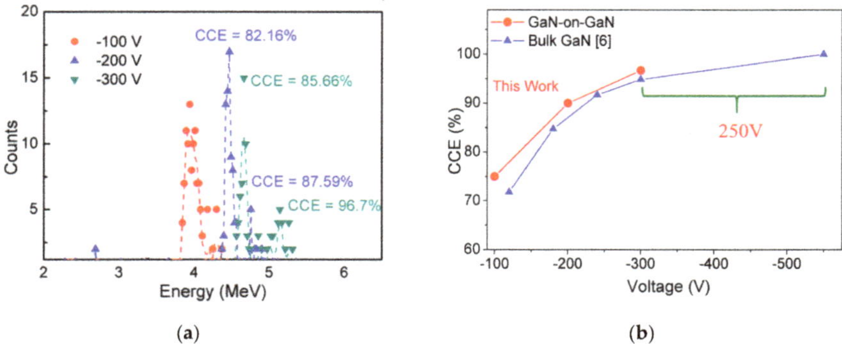

Figure 10. (**a**) Acquired α-particle spectra of GaN SBDs for different applied voltages (−100 V to −300 V) and (**b**) Comparison of measured CCE of SBDs vs. applied voltages (−100 V to −550 V) with state-of-the-art reported values.

The detector's energy resolution was extracted to be 71 keV from the full wave at half maximum (FWHM) of the α-particle spectra measured at −100 V. The measured energy resolution is 30% better than other reported bulk GaN-based detectors (121 keV) [6]. The improved energy resolution was due to the reduction in the straggling (statistical distribution of energy losses) of α-particles by placing the detector normal to the source. SRIM simulations were also performed to study the effect of the incident angle on energy resolution [27] (see Figure 11). An increase in the angle between the incident α-particle and the surface of the detector increases the FWHM of the detected spectral energy from 97.8 keV to 163.8 keV. This increase in FWHM is due to the increase in straggling when the source is placed at an angle to the detector, resulting in a reduction of straggling.

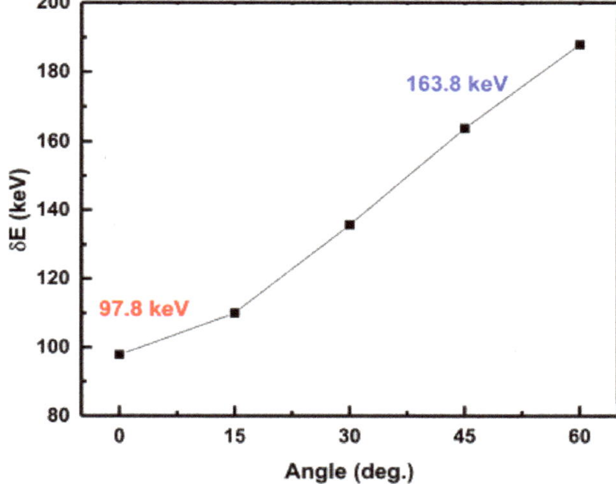

Figure 11. Variation of energy resolution with a change of the incident angle.

4.4. Benchmarking

Figure 12 compares the low voltage performance of GaN-on-GaN SBD detectors with the state-of-the-art α-particle detectors reported so far, as a function of detected energies. About 30% higher CCE was observed in the compensated 15 μm thick GaN DL-based α-particle detectors at −20 V. In addition, our detectors also exhibited 96.7% CCE at −300 V, which is 250 V lower than the published literature. These promising results pave the way to achieve high CCE, low operating voltage and portable α-particle detectors.

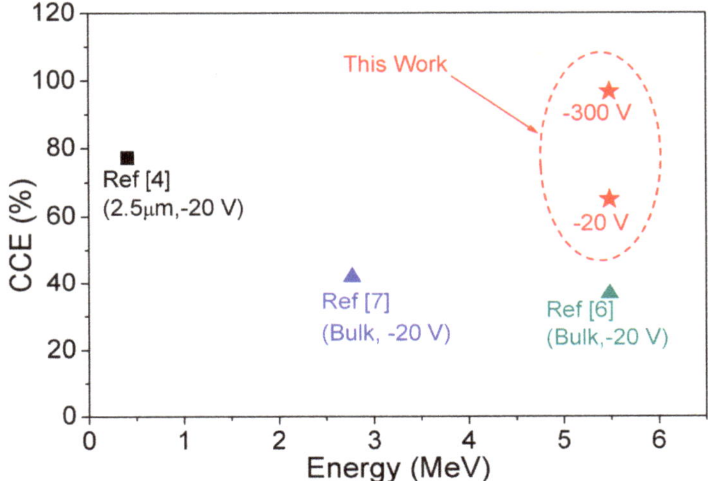

Figure 12. Benchmarking of extracted CCE of our detectors with epitaxial-grown GaN detectors (squares) and bulk GaN detectors (triangles) at low voltages.

5. Conclusions

In conclusion, we demonstrated a low-voltage (−20 V) operating 5.48 MeV α-particle detector with a record-high CCE of 65% using 15 μm thick compensated MOVPE-grown GaN DL on HVPE-grown bulk n+-GaN substrate. The measured CCE was 30% higher than the previously reported values at −20 V. The detectors also exhibited a high CCE of 96.7% at −300 V and the spectral resolution of 71 keV, which was 250 V lower and 30% better than the previously reported values. The improved performance in α-particle detection was due to the formation of a thicker DW, even at low voltages. The demonstrated vertical GaN-on-GaN SBD with compensated DL for portable α-particle detectors presented great potential to work even at low voltages.

Author Contributions: A.S. along with S.A. and G.I.N. have designed the experiments; Simulations and fabrication of devices was completed by A.S. and K.R.; Result analysis was completed by A.S., S.A. and G.I.N.; P.P.M. and J.K. performed radiation detection testing and analysis.; S.N., M.D., Y.H. and H.A. have grown the GaN material for fabrication of devices; A.S. wrote the paper, which was reviewed by all authors.

Funding: This research received no external funding.

Acknowledgments: The authors would like to express their gratitude to the staff at the nano-fabrication center in School of Electrical and Electronics Engineering, Nanyang Technological University for their support. We acknowledge Chris R. Purcell at GNS Science for his technical assistance during α-particle measurement.

Conflicts of Interest: The authors declare no conflict of interest.

References

1. Ionascut-Nedelcescu, A.; Carlone, C.; Houdayer, A.; von Bardeleben, H.J.; Cantin, J.L.; Raymond, S. Radiation hardness of gallium nitride. *IEEE Trans. Nucl. Sci.* **2002**, *49*, 2733–2738.
2. Weaver, B.D.; Anderson, T.J.; Koehler, A.D.; Greenlee, J.D.; Hite, J.K.; Shahin, D.I.; Kub, F.J.; Hobart, K.D. On the Radiation Tolerance of AlGaN/GaN HEMTs. *ECS J. Solid State Sci. Tech.* **2016**, *5*, Q208–Q212.
3. Eastman, L.F.; Mishra, U.K. The Toughest Transistor Yet. *IEEE Spectr.* **2002**, *39*, 28–33.
4. Vaitkus, J.; Cunningham, W.; Gaubas, E.; Rahman, M.; Sakai, S.; Smith, K.M.; Wang, T. Semi-insulating GaN and its evaluation for α particle detection. *Nucl. Inst. Methods Phys. Res. A* **2003**, *509*, 60–64.
5. Polyakov, A.Y.; Smirnov, N.B.; Govorkov, A.V.; Markov, A.V.; Kozhukhova, E.A.; Gazizov, I.M.; Kolin, N.G.; Merkurisov, D.I.; Boiko, V.M.; Korulin, A.V.; et al. Alpha particle detection with GaN Schottky diodes. *J. Appl. Phys.* **2009**, *106*, 103708.
6. Xu, Q.; Mulligan, P.; Wang, J.; Chuirazzi, W.; Cao, L. Bulk GaN alpha-particle detector with large depletion region and improved energy resolution. *Nucl. Inst. Methods Phys. Res. A* **2017**, *849*, 11–15.
7. Lee, I.H.; Polyakov, A.Y.; Smirnov, N.B.; Govorkov, A.V.; Kozhukhova, E.A.; Zaletin, V.M.; Gazizov, I.M.; Kolin, N.G.; Pearton, S.J. Electrical properties and radiation detector performance of free-standing bulk n-GaN. *J. Vac. Sci. Tech. B* **2012**, *30*, 021205.
8. Mulligan, P.; Wang, J.; Cao, L. Evaluation of freestanding GaN as an alpha and neutron detector. *Nucl. Inst. Methods Phys. Res. A* **2013**, *719*, 13–16.
9. Wang, G.; Fu, K.; Yao, C.S.; Su, D.; Zhang, G.G.; Wang, J.Y.; Lu, M. GaN-based PIN alpha particle detectors. *Nucl. Inst. Methods Phys. Res. A* **2012**, *663*, 10–13.
10. Grant, J.; Bates, R.; Cunningham, W.; Blue, A.; Melone, J.; McEwan, F.; Vaitkus, J.; Gaubas, E.; O'Shea, V. GaN as a radiation hard particle detector. *Nucl. Inst. Methods Phys. Res. A* **2007**, *576*, 60–65.
11. Zhu, Z.; Zhang, H.; Liang, H.; Tang, B.; Peng, X.; Liu, J.; Yang, C.; Xia, X.; Tao, P.; Shen, R.; et al. High-temperature performance of gallium-nitride-based pin alpha-particle detectors grown on sapphire substrates. *Nucl. Inst. Methods Phys. Res. A* **2018**, *893*, 39–42.
12. Lu, M.; Zhang, G.G.; Fu, K.; Yu, G.H. Gallium Nitride Room Temperature α Particle Detectors. *Chin. Phys. Lett.* **2010**, *27*, 052901.
13. Pullia, A.; Bertuccio, G.; Maiocchi, D.; Caccia, S.; Zocca, F. High-resolution alpha-particle spectroscopy with an hybrid SiC/GaN detector/front-end detection system. In Proceedings of the IEEE Nuclear Science Symposium Conference Record, Honolulu, HI, USA, 26 October–3 November 2007; pp. 516–520.
14. Melton, A.G. Development of wide bandgap solid-state neutron detectors. Ph.D. Thesis, Georgia Institute of Technology, Atlanta, GA, USA, May 2011.
15. Wang, J.; Mulligan, P.; Brillson, L.; Cao, L.R. Review of using gallium nitride for ionizing radiation detection. *Appl. Phys. Rev.* **2015**, *2*, 031102.
16. Polyakov, A.Y.; Smirnov, N.B.; Govorkov, A.V.; Markov, A.V.; Pearton, S.J.; Kolin, N.G.; Merkurisov, D.I.; Boiko, V.M. Neutron irradiation effects on electrical properties and deep-level spectra in undoped n- Al Ga N/Ga N heterostructures. *J. Appl. Phys.* **2005**, *98*, 033529.
17. Lin, C.H.; Katz, E.J.; Qiu, J.; Zhang, Z.; Mishra, U.K.; Cao, L.; Brillson, L.J. Neutron irradiation effects on gallium nitride-based Schottky diodes. *Appl. Phys. Lett.* **2013**, *103*, 162106.
18. Wang, L.; Mohammed, F.M.; Adesida, I. Differences in the reaction kinetics and contact formation mechanisms of annealed Ti/Al/Mo/Au Ohmic contacts on n GaN and AlGaN/GaN epilayers. *J. Appl. Phys.* **2007**, *101*, 013702.
19. Feng, Q.; Li, L.M.; Hao, Y.; Ni, J.Y.; Zhang, J.C. The improvement of ohmic contact of Ti/Al/Ni/Au to AlGaN/GaN HEMT by multi step annealing method. *Solid-State Electron.* **2009**, *53*, 955–958.
20. Hwang, Y.H.; Ahn, S.; Dong, C.; Zhu, W.; Kim, B.J.; Ren, F.; Lind, A.G.; Jones, K.S.; Pearton, S.J.; Kravchenko, I. Effect of Buffer Oxide Etchant (BOE) on Ti/Al/Ni/Au Ohmic Contacts for AlGaN/GaN Based HEMT. *ECS Trans.* **2015**, *69*, 111–118.
21. Yang, L. Planar and non gold metal stacks processes and conduction mechanism for AlGaN/GaN HEMTs on Si. Ph.D. Thesis, Nanyang Technological University, Singapore, 2016.
22. Huang, Y.; Huang, Z.; Zhong, Z.; Yang, X.; Hong, Q.; Wang, H.; Huang, S.; Gao, N.; Chen, X.; Cai, D.; et al. Highly transparent light emitting diodes on graphene encapsulated Cu nanowires network. *Sci. Rep.* **2018**, *8*, 13721.

23. Sandupatla, A.; Arulkumaran, S.; Ng, G.I.; Ranjan, K.; Deki, M.; Nitta, S.; Honda, Y.; Amano, H. Enhanced breakdown voltage in vertical Schottky diodes on compensated GaN drift layer grown on free-standing GaN. In Proceedings of the 13th International Conference on Nitride Semiconductors 2019 (ICNS-13), Bellevue, WA, USA, 7–12 July 2019.
24. Sandupatla, A.; Arulkumaran, S.; Ng, G.I.; Ranjan, K.; Deki, M.; Nitta, S.; Honda, Y.; Amano, H. Effect of GaN drift-layer thicknesses in vertical Schottky Barrier Diodes on free-standing GaN substrates. In Proceedings of the 2018 International Conference on Solid State Devices and Materials (SSDM2018), Tokyo, Japan, 19–22 September 2018.
25. Sandupatla, A.; Arulkumaran, S.; Ng, G.I.; Ranjan, K.; Deki, M.; Nitta, S.; Honda, Y.; Amano, H. GaN drift-layer thickness effects in vertical Schottky barrier diodes on free-standing HVPE GaN substrates. *AIP Adv.* **2019**, *9*, 045007.
26. Hu, Z.; Nomoto, K.; Song, B.; Zhu, M.; Qi, M.; Pan, M.; Gao, X.; Protasenko, V.; Jena, D.; Xing, H.G. Near unity ideality factor and Shockley-Read-Hall lifetime in GaN-on-GaN p-n diodes with avalanche breakdown. *Appl. Phys. Lett.* **2015**, *107*, 243501.
27. Tanaka, N.; Hasegawa, K.; Yasunishi, K.; Murakami, N.; Oka, T. 50 A vertical GaN Schottky barrier diode on a free-standing GaN substrate with blocking voltage of 790 V. *Appl. Phys. Exp.* **2015**, *8*, 071001.
28. Tompkins, R.P.; Khan, M.R.; Green, R.; Jones, K.A.; Leach, J.H. IVT measurements of GaN power Schottky diodes with drift layers grown by HVPE on HVPE GaN substrates. *J. Mater. Sci. Mater. Electron.* **2016**, *27*, 6108–6114.
29. Wang, J. Evaluation of GaN as a Radiation Detection Material. Master's Thesis, Ohio State University, Columbus, OH, USA, 2012.
30. Ziegler, J.F.; Ziegler, M.D.; Biersack, J.P. SRIM—The stopping and range of ions in matter (2010). *Nucl. Instrum. Methods Phys. Res. Sect. B* **2010**, *268*, 1818–1823.

Article

Effect of Commercial Off-The-Shelf MAPS on γ-Ray Ionizing Radiation Response to Different Integration Times and Gains

Shoulong Xu [1,2,*], Jaap Velthuis [3,4,], Qifan Wu [1,*], Yongchao Han [5,*], Kuicheng Lin [6], Lana Beck [3], Shuliang Zou [2], Yantao Qu [5] and Zengyan Li [7]

[1]   Department of Engineering Physics, Tsinghua University, Beijing 100084, China
[2]   School of resource Environment and Safety Engineering, University of South China, Hengyang 421001, China; zousl2013@126.com
[3]   School of Physics, University of Bristol, Bristol BS8 1QU, UK; Jaap.Velthuis@bristol.ac.uk (J.V.); lana.beck@bristol.ac.uk (L.B.)
[4]   School of Nuclear Science and Engineering, University of South China, Hengyang 421001, China
[5]   China Institute of Atomic Energy, Beijing 102413, China; quyantao@ciae.ac.cn
[6]   Instituted of materials, China academy of engineering physics, Mianyang 621700, China; linkc06@caep.cn
[7]   Northwest Institute of Nuclear Technology, Xi'an 710024, China; Lizy_THU@163.com
*   Correspondence: xushoulong@tsinghua.edu.cn (S.X.); wuqifan@tsinghua.edu.cn (Q.W.); hanyongchao@ciae.ac.cn (Y.H.); Tel.: +86-152-1181-2766 (S.X.)

Received: 21 October 2019; Accepted: 12 November 2019; Published: 13 November 2019

Abstract: We report the γ-ray ionizing radiation response of commercial off-the-shelf (COTS) monolithic active-pixel sensors (MAPS) with different integration times and gains. The distribution of the eight-bit two-dimensional matrix of MAPS output frame images was studied for different parameter settings and dose rates. We present the first results of the effects of these parameters on the response of the sensor and establish a linear relationship between the average response signal and radiation dose rate in the high-dose rate range. The results show that the distribution curves can be separated into three ranges. The first range is from 0 to 24, which generates the first significant low signal peak. The second range is from 25 to 250, which shows a smooth gradient change with different integration times, gains, and dose rates. The third range is from 251 to 255, where a final peak appears, which has a relationship with integral time, gain, and dose rate. The mean pixel value shows a linear dependence on the radiation dose rate, albeit with different calibration constants depending on the integration time and gain. Hence, MAPS can be used as a radiation monitoring device with good precision.

Keywords: COTS commercial MAPS; radiation response; integral time; gain

---

1. Introduction

In a nuclear accident or in a strong radiation field, the detection of high dose-rate radiation and wide-range γ-rays is expensive and inefficient [1]. The use of commercial off-the-shelf (COTS) complementary metal oxide semiconductor (CMOS) monolithic active-pixel sensors (MAPS) as a γ-ray radiation detector has been reported. Martín Pérez et al. observed that the radiation response characteristic can be used for radiometric imaging [2], and that MAPS can be used to classify particles and the sensor is sensitive to soft X-rays [3]. Galimberti and Wang reported successful radiation detection using commercial off-the-shelf MAPS [4]. Ma et al. used Advanced RISC Machine (ARM) microcontrollers and ZigBee modules in combination with MAPS to detect low-energy radiation [5,6]. Arbor et al. reported a linear relationship between the MAPS radiation response signal, and the dose rate of γ-ray radiation field was proven [7]. Early reports on the application feasibility of using smartphones as a radiation measure detector have been published [8], Wei et al. also reported using mobile phones with MAPS cameras for radiation detection, which confirmed that after calibration, smartphones can be used as γ-ray measuring devices and for radiation safety control of high-level radioactive sources such as industrial radiography, γ-ray irradiation facilities, and medical treatment [9–11]. Another report focused on determining the heavy particle effect using an active-pixel sensor, which produced a significant radiation response from a single event [12]. However, few papers report the effect of setting the parameters of a MAPS video surveillance camera on the radiation response signal in the strong γ-ray radiation range. According to the MAPS working process, the integral time and gain of the set parameters most directly affect the radiation response signal.

In this study, we examined the distribution of eight-bit two-dimensional matrix of the MAPS output frame image using different setting parameters and dose rates. The image data are expressed in the form of a distribution curve. The abscissa is the pixel value, and the ordinate is the count of pixels of the pixel value in the image. We present the first result of the effect of different parameter values on the response signal and the linear relationship between the statistical value of the response signal and radiation dose rate in the high-dose rate range.

The rest of this paper is organized as follows: The experimental setup and the data processing methods are detailed in Section 2. The experimental results and the data processing are described in Section 3. Our conclusions are presented in Section 4.

2. Experiments

2.1. Initial Parameters of Image Sensors

In the experiments, we used COTS CMOS MAPS (IMX 322 LQJ, Sony Corporation, Tokyo, Japan,) with 6.4 mm pixel-type and approximately 2.43 M effective pixels, which are low cost (about USD $35), have low background noise, and are easy to obtain. The sensor provides an 8-bit response for each pixel. The chips are operated with 2.7 V, 1.2 V digital, and 1.8 V interface triple power supplies. All the COTS sensors were shaded with a layer of opaque plastic material and placed inside a homemade dark box to prevent any interference from visible light and to maintain constant room temperature during video recording. The test sensor was employed with no lenses and was operated in monochromatic mode with a rolling shutter. The integration time of the sensor is adjustable from 1/25 to 1/10,000 s. The parameter settings of the sensors were controlled by software, and video was recorded at a frame rate of 25 frames per second (fps) in a video format and streamed to hard disk. The integration time and gain of the sensors were manually adjusted; all the autoregulation and noise reduction functions and the white balance were turned off. Based on previous research results, radiation damage can be ignored when the radiation dose is less than 30 Gy [13] for the MAPS with a 4 T structure used in this study. The background noise with pixel values between 1 and 5 is the most common before irradiation, and the number of pixels with values greater than 10 is less than 0.2% of the total. The picture of the MAPS module samples is shown in Figure 1.

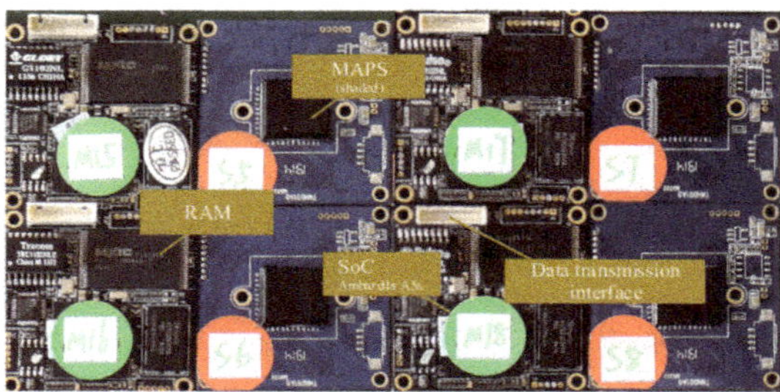

Figure 1. Picture of the monolithic active-pixel sensors (MAPS) module.

2.2. Experimental Setup

A cylindrical ^{60}Co source from the China Institute of Atomic Energy (CIAE, Beijing, China) was used in all experiments. ^{60}Co provides γ-rays of 1.17 and 1.33 MeV. The average activity of the source was 130 kCi, whereas the radiation non-uniformity was less than 5%. The ambient temperature during sensor testing was 20 °C. The CMOS MAPS sample sensors were irradiated with dose rates ranging from 51.61 to 479.24 Gy/h, which were obtained from low to high dose rates using a movable slider. The dose rates at each point on the slide track were calibrated. The total ionizing dose was measured using a radiochromic film dosimeter, and the dose rate was calculated as the ratio of the total ionizing dose to the irradiation time. The experimental setup is shown in Figure 2. The test sensors were operated nearly continuously during all experiments. Signal data were transmitted using a 4800LX cable, and the maximum video bitrate was 13 Mbps. Video files were recorded at 25 fps by a computer during all experiments, and the data were imported using MATLABR 2014a (Math Works Inc., Natick, MA, USA) and then split into individual frames. During experiments, the systems controlling aperture, shutter, gain, and white balance were set to manual, and noise reduction functions and exposure compensation functions were turned off. All the data were collected before the dose rate reached 30 Gy.

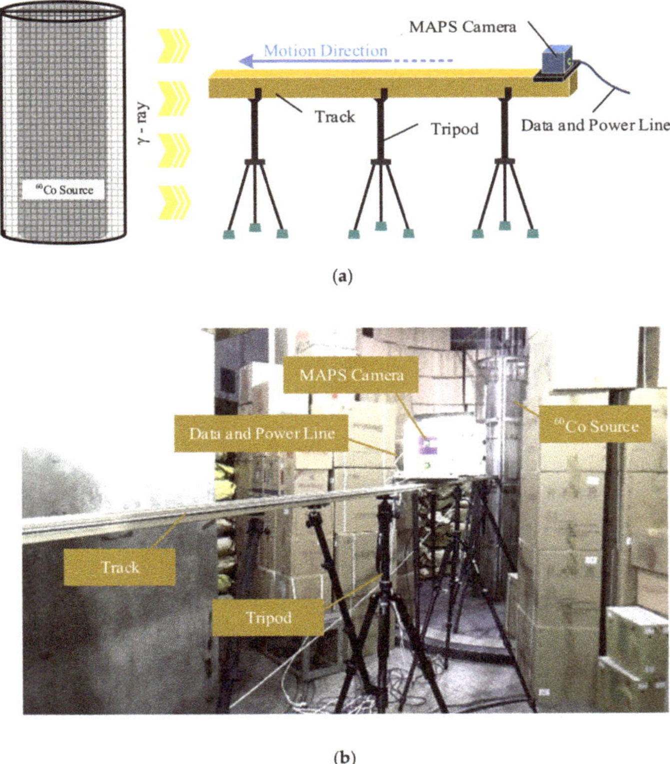

Figure 2. Experimental setup: (**a**) Schematic view of the experimental system and (**b**) picture of the experimental setup.

The recorded video data were processed on a PC by MATLAB. Each frame of the video was transformed into an 8-bit gray value matrix for analysis. The 8-bit gray value is the sensor in analog-to-digital converter (ADC) units, and the range of gray value was from 0 to 255. To obtain more accurate statistics, radiation response events in 100 consecutive frames were counted together. The mean pixel value (S_k) of the selected images at the radiation dose rate of k was calculated as follows:

$$S_k = \frac{1}{MN} \sum_{j=1}^{j=M} \sum_{i=1}^{i=N} \left(E_{i,j} - I_{i,j} \right) \tag{1}$$

where $I_{i,j}$ is the pixel value of the ith pixel in the jth frame before irradiation, $E_{i,j}$ is the pixel value of the ith pixel in the jth frame at the radiation dose rate of k, M is the number of frames, and N is the pixel count.

3. Discussion

Figure 3 shows the distribution of the count fraction of pixel signals at a dose rate of 51.61 Gy/h and using two gains, 6 and 42 dB, in frames captured with integration times ranging from 1/8000 to 1/25 s. For a gain of 6 dB (Figure 3a), the first peak corresponds to gray levels below 15. The position of peak shifts to larger pixel values with increasing of integration time, which indicates that more pixels yield stronger pixel signals due to exposure to more photons. However, no obvious change in the position of peak for the larger gain of 42 dB (Figure 3b) was observed. For both amplifications, the height of the pixel value distributions between 75 and 200 followed the integration time; longer integration times yield higher count fraction distributions. The count and maximum value of peaks in that range increased with larger integration time, and curves at 42 dB were smoother than those at 6 dB with the exception of integration times of 1/100 and 1/240 s. This indicates that the distribution for lower integration times and larger gains is smoother, which indicates that a quantization issue exists in the sensor. We noticed that a significant peak exists in the range larger than 250; the peak is narrower and smaller at the larger gain of 42 dB. The shape of radiation response events has been reported [13], and this peak might be caused by some saturation and supersaturation radiation response events in frames.

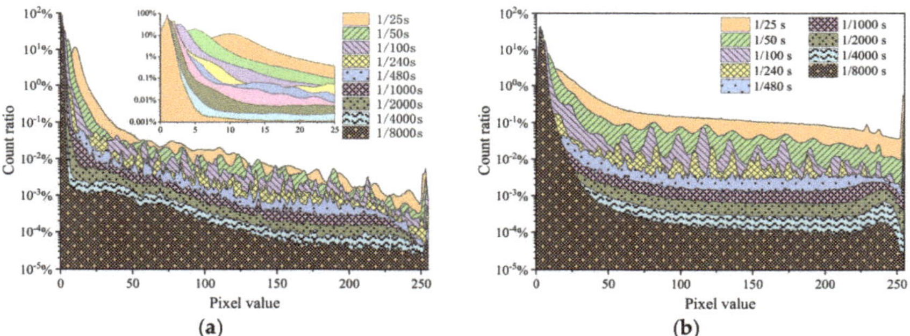

Figure 3. Distribution of the count fraction of pixels at a dose rate of 51.61 Gy/h and (**a**) a gain of 6 dB and (**b**) a gain of 42 dB in frames captured at using integration times ranging from 1/8000 to 1/25 s.

We calculated the average pixel value in Figure 3 in the range between 75 to 250 for each integration time. The relationship between the average pixel value and integration time at the irradiation dose rate of 51.61 Gy/h is plotted in Figure 4. Figure 4 shows that for integration times larger than 1/480 s, a linear relationship exists between the average pixel value and integration time. The linearity for 6 dB is better than that of 42 dB.

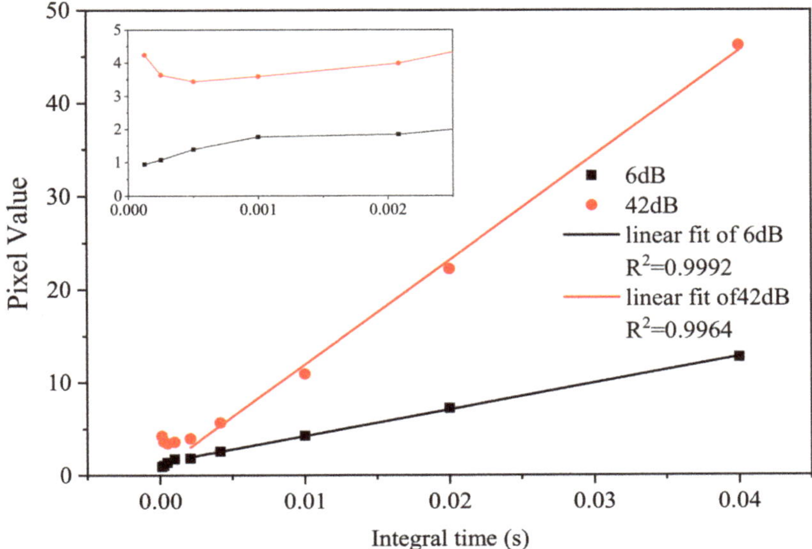

Figure 4. Average pixel value as a function of the integration time for a gain of 6 and 42 dB during irradiation at 51.61 Gy/h.

The distribution of count fractions of pixels at a gain of 6 dB and at six irradiation dose rates from 64.48 to 265.22 Gy/h at three integration times of 1/100, 1/240, 1/480 s are shown in Figure 5. The pixel value was calculated using Equation (1). As the irradiation dose rate increases, the maximum value increases, and the number of pixels with values larger than 25 increases. A peak for pixel signals larger than 250 occurred at irradiation doses rate larger than 200 Gy/h. This peak occurred at all measured integration times.

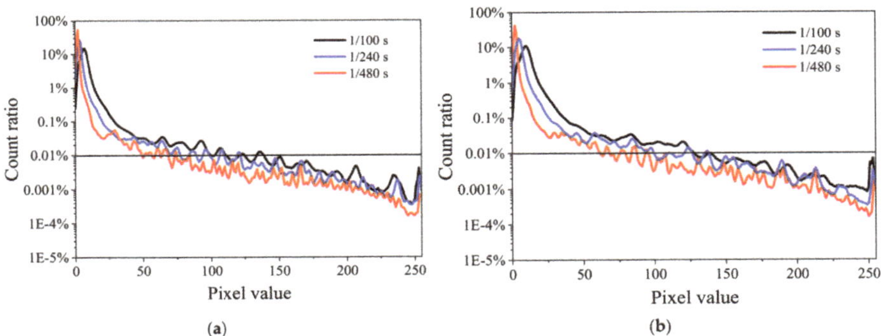

(a)   (b)

Figure 5. *Cont.*

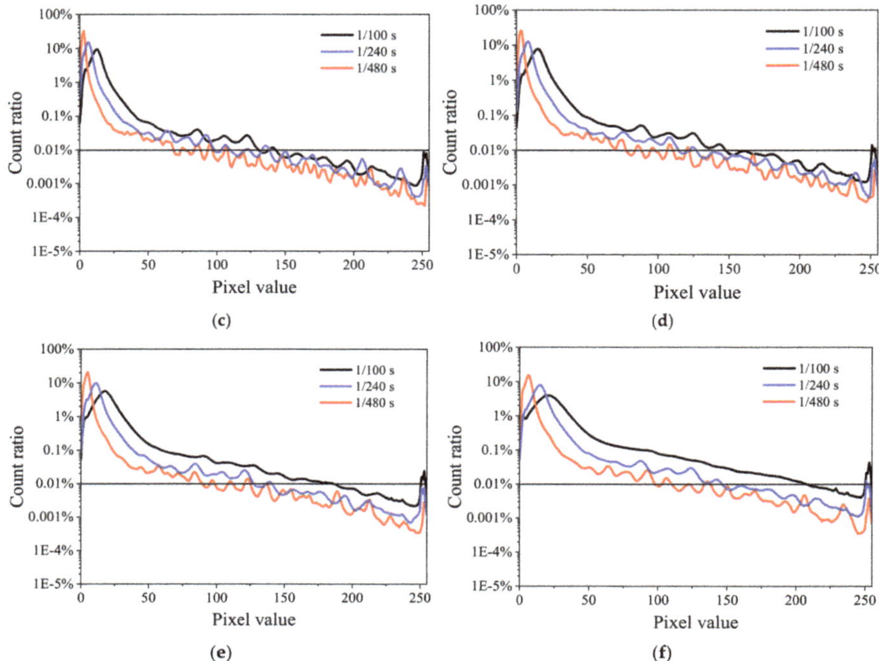

Figure 5. Distribution of the count fractions of pixels at a gain of 6 dB for six different dose rates in frames captured at integration times of 1/100, 1/240, and 1/480 s: (**a**) 64.48, (**b**) 95.00, (**c**) 119.5, (**d**) 153.41, (**e**) 200.32, and (**f**) 265.22 Gy/h.

We calculated the average pixel signal in Figure 5 in the range between 25 to 250, i.e., excluding the peak around 250. The relationship between mean pixel value and dose rate at the integration times of 1/100, 1/240, and 1/480 s is shown in Figure 6. The linearities of the fit of 1/100, 1/240, and 1/480 s are 0.9985, 0.9986, and 0.9964, respectively.

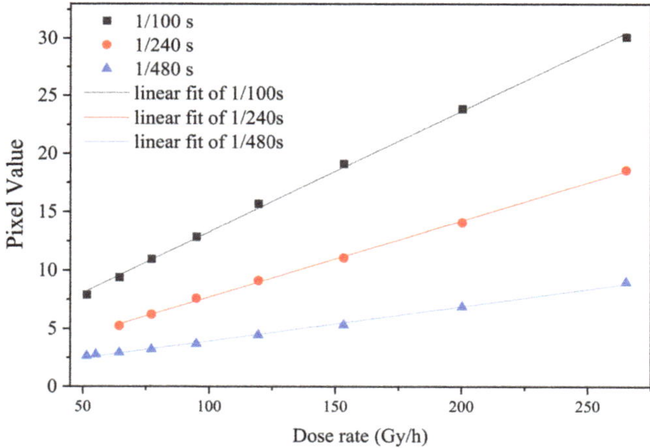

Figure 6. Average pixel signal as a function of irradiation dose rate for integration times of 1/100, 1/240, and 1/480 s at a gain of 6 dB.

Figure 7 shows the distribution of count fractions of pixels at a gain of 6 dB and integration times of 1/240 and 1/480 s at dose rates ranging from 64.48 to 265.22 Gy/h. The pixel value was calculated using Equation (1). The position of the peak shifts to larger pixel values with increasing dose rates as more pixels receive hits and even multiple hits. This is particularly clear for the first peak, i.e., the peak with values below 25. Figure 7 compares the effects of dose rate changes on the distribution under different integration times. The peak locations of these curves in the range larger than 25 also depend on the integration time. The same peak structure was observed for both integration times, but the peak locations move with integration time.

Figure 8 shows the distribution of the count fraction of pixels at an integration time of 1/100 s at four dose rates captured using gains of 6, 12, 24, and 42 dB. The figure compares the effect of adjusting the gain on the statistical curve under different dose rates. The graphs show that for higher gains, more pixels have higher signals, and higher radiation dose rates yield higher pixel values. All graphs display peaks in the same location irrespective of the gain. This indicates issues with the sensor.

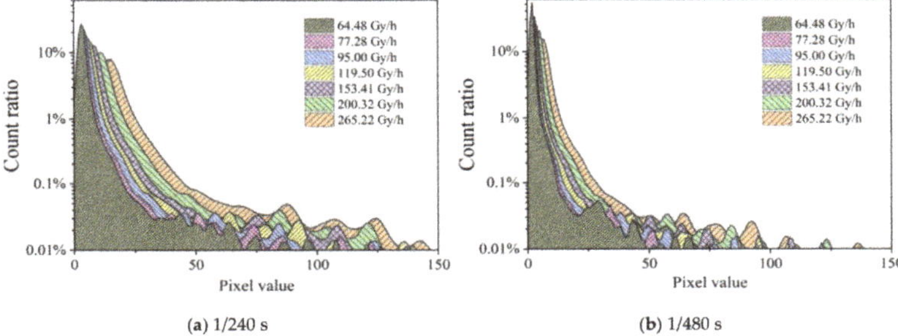

Figure 7. Distribution of the count fraction of pixels at a gain of 6 dB and for two integration times captured at dose rates ranging from 64.48 to 265.22 Gy/h.

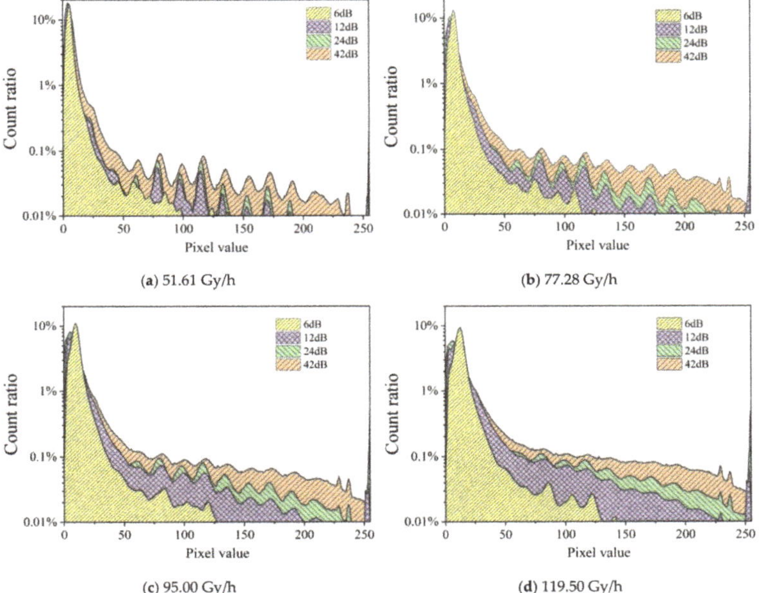

Figure 8. Distribution of the count fraction of pixels for an integration time of 1/100 s at four dose rates captured using gains of 6, 12, 24, and 42 dB.

Figure 9 shows the distribution of the count fraction of pixels at an integral time of 1/100 s captured at different dose rates ranging from 51.61 to 119.50 Gy/h, and measured using gains of 12, 24, and 48 dB using dose rates ranging from 51.61 to 479.24. For 6 dB, we observed that the peak at low pixel values moves to the right with increasing dose rate. However, this was not observed for any of the higher gains. Figure 9 shows that larger gains result in more pixels with larger values. Peaks are generated at the same positions for gains from 12 dB. This indicates a quantization issue in the data as the same photons should yield higher signals at higher gains.

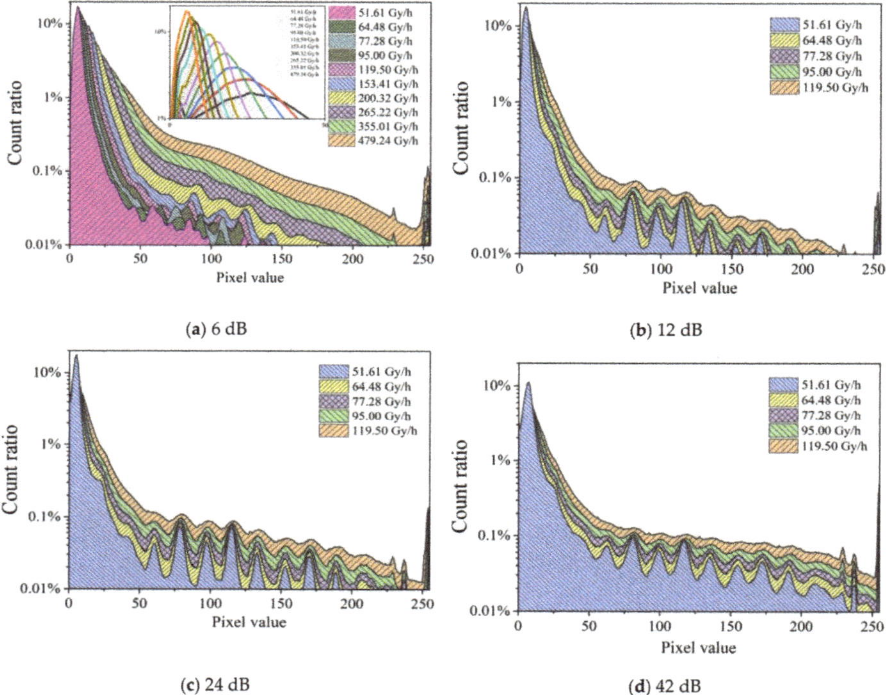

(a) 6 dB

(b) 12 dB

(c) 24 dB

(d) 42 dB

Figure 9. Distribution of the count fraction of pixels at an integration time of 1/100 s for four values of the gain captured at dose rates ranging from 51.61 to 119.50 Gy/h (from **b** to **d**), and from 51.61 to 479.24 Gy/h (**a**).

In summary, we can control the distribution range of pixel values between 0 and 255 by adjusting the integration time or gain and we can separate the distribution curves into three ranges. The first range is 0 to 24, which incorporates the first significant peak. The second range is 25 to 250, which shows a smooth gradient change with different integration times, gains, and dose rates. The last range is 251 to 255, where a peak occurs that is related to integration time, gain, and dose rate. Since the γ-ray dose rate detection relies on the pixel values of frames, studying the response signals is crucial. The results show that a more stable response is obtained for larger gains and lower integral times. However, a lower integral time means less sampling efficiency of the radiation response signal and a smaller dynamic range, which is an important factor affecting the detection accuracy and efficiency.

Figure 10 shows the dependence of the mean pixel signal on the irradiation dose rate for gains of 6, 12, 24, and 42 dB at an integration time of 1/25 s, where only pixels with signals ranging between 25 to 250 in Figure 9 are included. The pixel value was calculated using Equation (1). The linearity of the linear fit of 6, 12, 24, and 42 dB are 0.9997, 0.9996, 0.9990, and 0.9998, respectively, demonstrating good linearity.

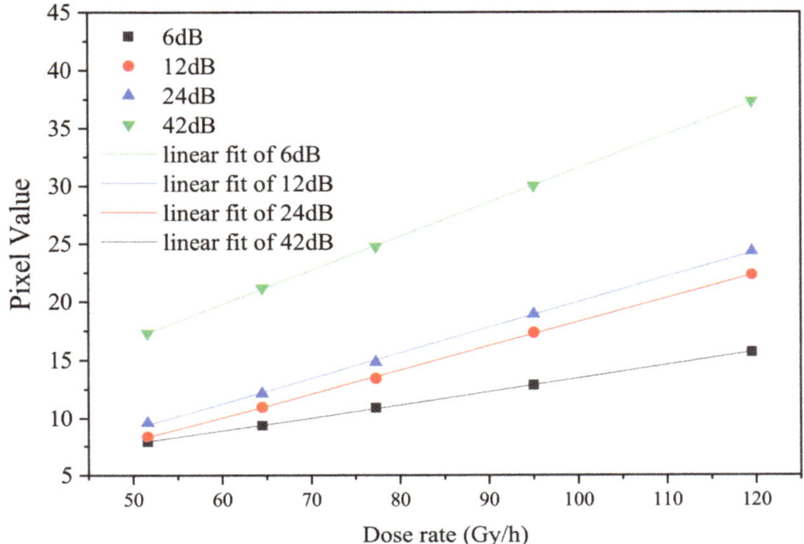

Figure 10. Mean pixel value as a function of the irradiation dose rate for gains of 6, 12, 24, and 42 dB at an integration time of 1/25 s.

4. Conclusions

In this work, we investigated the effect of integration time and gain of COTS MAPS on ionizing radiation detection. We discussed the potential use of MAPS at a high dose rate and wide range γ radiation detector. The pixel response distribution range of pixel value can be controlled by adjusting the integration time or gain from 50 to 265 Gy/h. The distribution curves can be separated into three ranges. The first range is 0 to 24, which generates the first significant peak. The second range is from 25 to 250, which shows a smooth gradient change with different integration times, gains, and dose rates. The last range is 251 to 255, which might be caused by some saturation and supersaturation radiation response events in the frames. This peak occurs for all measured integration times at an irradiation dose rate ranging from 51.61 to 265.22 Gy/h. More pixels with larger pixel values are generated with larger dose rates, which shifts the position of peak to larger pixel values. The mean pixel value shows a linear dependence on the radiation dose rate, albeit with different calibration constants depending on the integration time and gain. Hence, MAPS can be used with good precision as a radiation monitoring device under different settings for different doses.

Author Contributions: Conceptualization, S.X. and J.V.; Data curation, Y.H. and Z.L.; Formal analysis, Y.H. and L.B.; Funding acquisition, S.X., K.L., and Y.H.; Investigation, Q.W.; Methodology, S.X. and Y.H.; Software, S.X. and Q.W.; Supervision, S.Z. and Y.Q.; Writing—original draft, S.X. and L.B.; Writing—review & editing, S.X. and J.V.

Funding: This research was funded by the National Natural Science Foundation of China (Project No. 11905102 and Project No. 61601423). The experiments described here were funded by Hunan Provincial Key Laboratory of Emergency Safety Technology and Equipment for Nuclear Facilities and partly by the School of Nuclear Science and Technology of the University of South China.

Acknowledgments: The authors would like to express their sincere gratitude to the China Institute of Atomic Energy for providing the ^{60}Co γ source and the nuclear radiation detector.

Conflicts of Interest: The authors declare no conflict of interest.

References

1. Wang, X.; Zhang, P.; Liu, X. Calculation and Analysis for the Radiation Condition in the Containment of Pwr After Severe Accident. *Chin. J. Nucl. Sci. Eng.* **2013**, *33*, 163–167.
2. Alcalde Bessia, F.; Pérez, M.; Lipovetzky, J.; Gomez Berisso, M.; Piunno, N.; Mateos, H.; Pomiro, F.J.; Sidelnik, I.; Blostein, J.J.; Sofo Haro, M. X-ray Micrographic Imaging System Based on Cots Cmos Sensors. *Int. J. Circuit Theory Appl.* **2018**, *46*, 1848–1857. [CrossRef]
3. Haro, M.S.; Bessia, F.A.; Pérez, M.; Blostein, J.J.; Balmaceda, D.F.; Berisso, M.G.; Lipovetzky, J. Soft X-rays Spectroscopy with a Commercial Cmos Image Sensor at Room Temperature. *Radiat. Phys. Chem.* **2019**, 108354. [CrossRef]
4. Galimberti, C.L.; Bessia, F.A.; Perez, M.; Gomez Berisso, M.; Haro, M.S.; Sidelnik, I.; Blostein, J.; Asorey, H.; Lipovetzky, J. A Low Cost Environmental Ionizing Radiation Detector Based on Cots Cmos Image Sensors. In Proceedings of the 2018 IEEE Biennial Congress of Argentina, San Miguel de Tucumán, San Miguel de Tucuman, Argentina, 6–8 June 2018; pp. 1–6.
5. Wang, X.; Zhang, S.L.; Song, G.X.; Guo, D.F.; Ma, C.W.; Wang, F. Remote measurement of low-energy radiation based on ARM board and ZigBee wireless communication. *Nucl. Sci. Tech.* **2018**, *29*, 4. [CrossRef]
6. Chen, Q.Q.; Yuan, Y.Z.; Ma, C.W.; Wang, F. Gamma Measurement Based on Cmos Sensor and Arm Microcontroller. *Nucl. Sci. Tech.* **2017**, *28*, 122. [CrossRef]
7. Arbor, N.; Higueret, S.; Elazhar, H.; Combe, R.; Meyer, P.; Dehaynin, N.; Taupin, F.; Husson, D. Real-time Detection of Fast and Thermal Neutrons in Radiotherapy with CMOS Sensors. *Phys. Med. Biol.* **2017**, *62*, 1920–1934. [CrossRef] [PubMed]
8. Pang, L.Y. Can Smartphones Measure Radiation Exposures? *Pac. Hist. Rev.* **2015**, *58*, 387–388. [CrossRef]
9. Wei, Q.Y.; Bai, R.; Wang, Z.P.; Yao, R.-T.; Gu, Y.; Dai, T.-T. Surveying Ionizing Radiations in Real Time Using a Smartphone. *Nucl. Sci. Tech.* **2017**, *28*, 70. [CrossRef]
10. Tith, S.; Chankow, N. Measurement of Gamma-rays Using Smartphones. *Open J. Appl. Sci.* **2016**, *6*, 31–37. [CrossRef]
11. Kang, H.G.; Song, J.J.; Lee, K.; Nam, K.C.; Hong, S.J.; Kim, H.C. An Investigation of Medical Radiation Detection Using Cmos Image Sensors in Smartphones. *Nucl. Instrum. Methods Phys. Res.* **2016**, *823*, 126–134. [CrossRef]
12. Cai, Y.; Guo, Q.; Li, Y.; Wen, L.; Zhou, D.; Feng, J.; Ma, L.-D.; Zhang, X.; Wang, T.-H.; Zhang, X. Heavy Ion-induced Single Event Effects in Active Pixel Sensor Array. *Solid State Electron.* **2019**, *152*, 93–99. [CrossRef]
13. Xu, S.; Zou, S.; Han, Y.; Zhang, T.; Qu, Y.; Shoulong, X.; Shuliang, Z.; Yongchao, H.; Taoyi, Z.; Yantao, Q. Obtaining High-Dose-Rate gamma-Ray Detection With Commercial Off-the-Shelf CMOS Pixel Sensor Module. *IEEE Sens. J.* **2019**, *19*, 6729–6735. [CrossRef]

Article

Using a Scintillation Detector to Detect Partial Discharges

Łukasz Nagi *, Michał Kozioł, Michał Kunicki and Daria Wotzka

Institute of Electrical Power Engineering and Renewable Energy, Opole University of Technology, Prószkowska 76 Street, 45–758 Opole, Poland; m.koziol@po.edu.pl (M.K.); m.kunicki@po.edu.pl (M.K.); d.wotzka@po.edu.pl (D.W.)
* Correspondence: l.nagi@po.edu.pl

Received: 7 October 2019; Accepted: 11 November 2019; Published: 13 November 2019

Abstract: This article presents the possibility of using a scintillation detector to detect partial discharges (PD) and presents the results of multi-variant studies of high-energy ionizing generated by PD in air. Based on the achieved results, it was stated that despite a high sensitivity of the applied detector, the accompanying electromagnetic radiation from the visible light, UV, and high-energy ionizing radiation can be recorded by both spectroscopes and a system commonly used to detect radiation. It is also important that the scintillation detector identifies a specific location where dangerous electrical discharges and where the E-M radiation energy that accompanies PD are generated. This provides a quick and non-invasive way to detect damage in insulation at an early stage when it is not visible from the outside. In places where different radiation detectors are often used due to safety regulations, such as power plants or nuclear laboratories, it is also possible to use a scintillation detector to identify that the recorded radiation comes from damaged insulation and is not the result of a failure.

Keywords: radiation sensing technologies; partial discharges; scintillations; air insulation; photomultiplier

1. Introduction

The earlier a source of partial discharges (PD) is detected, the faster it can be reacted to, thus reducing the probability and cost of failure. PD detection is the recording of signals from phenomena associated with discharges. Early detection of electrical discharges in electrical devices and wires is an important part of diagnostics in electrical engineering. The subject area of this paper relates to high-voltage diagnostics with respect to PD. Understanding the phenomenon of electric discharges, their occurrence mechanisms, and propagation is a significant factor that helps to improve the diagnostics of energy devices. Research studies associated with electrical discharges increase the overall knowledge of this dangerous phenomenon and contribute to the development of the branch. Currently, in the high-voltage diagnostics of insulation systems, several methods are applied, which allow for the assessment and evaluation of PD. The most widely-known and commonly used research methods are the electric [1] and Ultra High Frequency (UHF) [2–5] methods, dissolved gas analysis (DGA) or Density Funcitonal Theory (DFT) [6,7] method, and the acoustic emission method [8,9]. During the occurrence of PD, air ionization and ozone production occur. Ozone testing can be effective, in itself, to assess PD levels [10]. The fact that the oxygen molecules are ionized shows the appearance of high-energy E-M radiation. Technological progress in optics enabled the development of a relatively new method for identification and localization of PD in air, with regards to the spectroscopy method [11,12]. Spectrophotometer enables for a more precise determination of the location where PD occurs. However, there is a gap in research on PD, both in the energy balance and in the phenomena related to the generation of PD. There are studies suggesting the generation of electromagnetic radiation during PD [13,14]. They can have a significant impact both on energy loss and safety, but also on the results of measurements of other values measured during PD detection.

They may also have an impact on faster wear of electrical insulation in cables or on faster degradation of the cable itself. Some articles describe the stimulation to generate PD with X-ray pulses [15,16]. The discharges themselves generate X-rays. All of this leads to the destruction of the insulation. It can also cause a false interpretation of the E-M radiation source in objects such as nuclear laboratories and nuclear power plants. In such cases, it is worth knowing that the source of the registered radiation is PD. Some articles describe PD that occur in fission chambers [17]. This article and the studies presented in it fill this gap. Additionally, with regard to the influence of radiation on the conductors and their insulation.

In this paper, we consider the emission of high-energy electromagnetic radiation. The main aim is to investigate if and how much high-energy radiation is emitted by PD. Based on previous studies, it was stated that emission of high-energy ionizing radiation constitutes a significant phenomenon occurring during the generation of PD in oil.

The condition of the electric discharge is the presence of ionizing factors or free-electron sources. In gases, e.g., the air, which is a natural dielectric, light flashes and other accompanying effects, e.g., acoustic, can be observed. Other phenomena, which are not immediately visible, can be measured using many methods. Often, the discharge is determined by the phenomena resulting from the acceleration of electrons and ions by the electric field in the discharge itself.

Complete discharges may occur as a spark or electric arc. This form causes a low resistance short circuit of the electrodes. The discharge occurs between two electrodes and is characterized by low internal electrical resistance. In the area of the electric arc, the gas is highly ionized and reaches a high temperature. In the air, under atmospheric pressure and with a current flow of 1 Ampere, it is about 5000–6000 Kelvin. The ionized gas is in the form of a plasma and its temperature depends on the type of gas, its pressure, but also on the current, and the type of electrodes. A PD is a local breakdown of a small part of the electrical insulation under high-voltage stress.

The mechanism of PD and the process of breakdown using gas insulation are similar to those in ideal liquid dielectrics. Insulating liquids in real systems are often contaminated with dissolved gasses or small solids. All of this influences the process of dielectric decomposition. Air testing is an attempt to measure the energy of the radiation itself in a facility where it is easier to register. The results can be used in subsequent studies to model the phenomenon in different isolation centers or under different environmental conditions.

Scintillation detectors are widely used in medical and industrial imaging. Different scintillation materials and photomultiplier models are used in the construction. Choosing the right scintillator and photomultiplier for planned research and measurements requires determining the characteristics of the scintillation material. New materials are still being produced [18,19]. Existing solutions in different configurations are also being tested [20]. Reducing the impact of radiation on electrical equipment can be achieved by selecting the right scintillator. The most widely used scintillator is NaI(Tl). Its properties have been studied since the 1950s. Recently, its spectroscopic and luminescent properties at liquid nitrogen temperature have also been studied [21]. Apart from the scintillator, the detection systems also have detectors of light coming from the scintillator. The most commonly used are conventional photomultipliers (PMT). More and more often, however, detectors using silicon photomultiplier (SiPM) are used. The energy resolution for the NaI-SiPM detector is as good as the NaI-PMT detector [22].

2. Measuring Setup

The main object in the test stand is to create a system for generating full and partial electrical discharges. It is comprised of a set of PD model sources supplied by a high-voltage transformer. The transformer is controlled from a control panel, which is connected to a computer. In order to reduce interference of the electric field, a wireless driver for Xbee radio link is applied for data transmission. The component applied for electromagnetic radiation measurement was made up of a scintillation detector based on a scintillating crystal, photomultiplier tube, a measuring card (Figure 1), and a compact precise controlled 3-dimensional (CPC3D) system, which allows for a precise positioning of the detector in any point in space with an accuracy of 0.1 mm. As it was written, the radiation measurement system consisted of a scintillation detector. The detector itself can be divided into three parts. The first element is a scintillation crystal responsible for changing high-energy radiation to radiation with a different wavelength range. Another element of the detector is a photomultiplier. The last element is a measuring card. The scintillation crystal NaI(Tl) with dimensions of 1.5″ × 1.5″ was used in the construction of the scintillation detector. The scintillator was placed in front of the photomultiplier in a scintillation probe directed towards the radiation source. In the measurements, the Adit photomultiplier, model B38B01W with the maximum voltage between the cathode and the anode being 1500 V, was used. The photomultiplier had a system of 6 dynodes amplifying the signal. The photocathode of the photomultiplier was a semitransparent alloy called Bialkali (Sb-K-Cs) and it cooperated well with the NaI(Tl) crystal. The spectral range of the photomultiplier operation ranges from 300 nm to almost 650 nm of visible light wavelength. The photomultiplier and scintillator consisted of a scintillation probe manufactured by Ludlum. The detectors power supply voltage was 700 V, while the preamplifier was supplied by a 3.7 V lithium-polymer battery with a capacity of 13,000 mAh. The preamplifier applied was the GS-1100 PRO from Gamma Spectacular. Data recorded by the detector and its measuring card are transmitted to the computer via fiber optic cables. It is possible thanks to the use of an integrated circuit, based on the PIC32 microcontroller and HFBR1521 module. The task of the system is to receive data from the measuring card and transmit it in real-time to the computer. For this purpose, the circuit board receives data from the measuring card, then they are modulated at the input to the fiber optic link, and then demodulated and transmitted to the computer. To convert the voltage scintillation pins into a digital form, the LTC1864 analog–digital converter manufactured by Linear Technology is used. The converter is a 16-bit chip with a maximum sampling frequency of 250 kHz. The power supply current shall not exceed 0.85 mA. The driver uses the successive approximation (SAR) method using switched capacities. Communication with the host microcontroller is via the serial interface SPI. The circuit is powered by 5V.

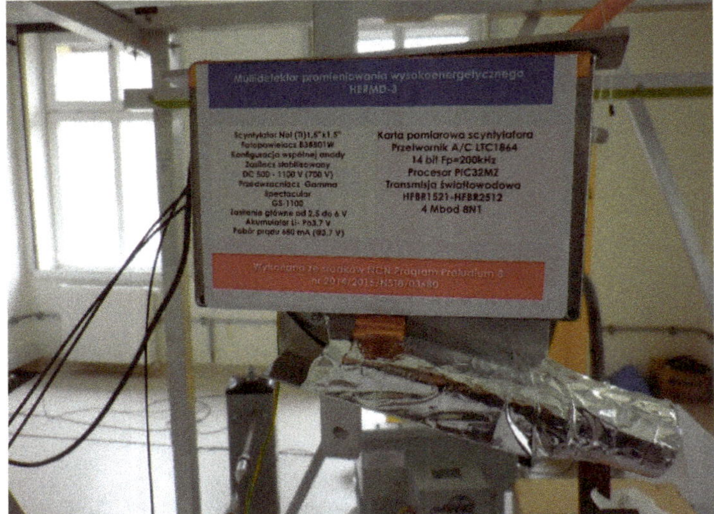

Figure 1. Scintillation detector based on a scintillating crystal, photomultiplier tube, and a measuring card.

The construction of the CPC3D is 2 × 2 × 2.5 m and is aluminum-based with an arm able to move in three dimensions. The detector is installed on the arm and is positioned next to the negative electrode during measurements, or at a specified distance from this electrode depending on a test trial.

Tests were performed for the following 3 systems enabling the generation of PD, a needle–needle system, a sphere–sphere system, and a system for generating surface discharges.

At the same time, for the possibility of verifying the obtained results, the PD phenomenon was also measured using the electric method. For apparent charge measurements, the MPD600 system is used, which meets all modern standards and is contemporary. The measurement configuration consisted of a 1 nF MCC210 coupling capacitor, a 30 µF four-pole CPL542A (also used for impedance measurement), and an MPD600 as a PD analyzer.

3. Measurement Methodology

Prior to the measurements, the detector was calibrated using a source of radiation, which was, in this case, americium (^{241}Am). Afterward, americium was placed near the electric field generated by surface discharges. The distance between the detector and the source of discharges was 30 cm, whereas americium was positioned between the detector and the positive electrode near the detector. During the calibration procedure, supply voltage value was increased until the breakdown of the system occurred (see Figure 2b). The energy radiation spectrum was recorded during the voltage increase period, several minutes after its disappearance, and subsequently several seconds after americium was removed from the detector area. The registered signals are shown in Figure 2a. From Figure 2a, energy radiation of significant quantity (up to 30 counts) and characteristic values (about 20 keV and 50 keV characteristic for ^{241}Am) can be recognized independently of the supply voltage value. After removal of the americium, no high-energy signals were detected. Both during calibration and radiation measurements, stable environmental conditions were ensured, i.e., constant temperature, humidity, and atmospheric pressure. A more detailed description of the measuring system and the calibration of the detectors can be found in [13,23].

As an insulating medium, air at atmospheric pressure was considered in the presented research studies. Tests were performed for 3 systems enabling generation of PD:

o   A needle–needle system, installed at various distances between the electrodes, changed with a step of 20 mm from 20 mm to 200 mm for each measurement trial;

o   A sphere–sphere system (the diameter of the spheres = 50 mm) for distances between the electrodes from 20 mm to 200 mm, increased by 20 mm for every next measurement;

o   A system for generating surface discharges, where a 3 mm glass plate was applied as the insulator between the electrodes.

Investigations of surface discharges were carried out for three different distances between the detector and the PD source, as well as for three different distances between the positive electrode and the glass plate (for each distance between the detector and the PD source). During the measurements, the supply voltage value was increased until the breakdown of the system occurred.

At the same time, together with radiation measurements, the electrical method was applied to determine the apparent charge of the phenomenon. The possibility to relate the results to existing and recognized methods is intended to demonstrate that radiation detection can be an important means of the non-invasive location of PD. It is also possible to verify that PD occurred during the recording of energy in the X-ray range. Examples of electrical measurement results obtained are presented in Table 1, where Q_{IEC} is the apparent charge compliant with the standard IEC 60270, Q_{Peak} is the maximum apparent charge, N is the number of counts of PD/s, I_{Dis} and P_{Dis} are the current and power of discharge, and D is the PD density. Results of measurements using electric method in application to similar electrode systems can be found in [3].

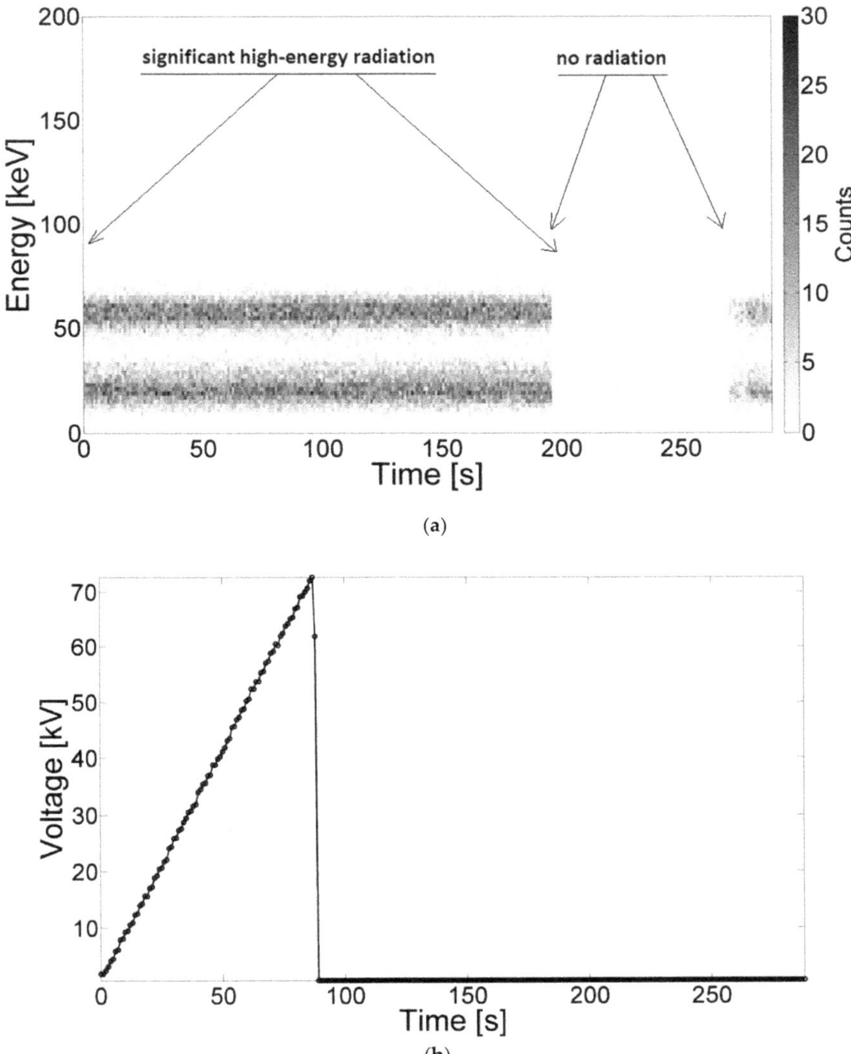

Figure 2. The energy spectrum of americium recorded by the detector (**a**) and a graph of increasing voltage between the electrodes during the test (**b**) until the breakdown.

Table 1. Examples of electrical method measurement.

Parameter	Needle–Needle System	Sphere–Sphere System
Breakdown voltage	72 kV	64 kV
Q_{IEC}	353.4 nC	18.57 nC
Q_{Peak}	504.7 nC	28.74 nC
N	57.19 kPD/s	68.14 kPD/s
I_{Dis}	1.261 mC/s	40.01 mC/s
P_{Dis}	105.7 W	2.354 W
D	167.4 pC^2/s	133.1 fC^2/s

4. Analysis of the Results

Examples of the energy spectra recorded during measurements in the needle–needle and sphere–sphere systems are depicted in Figures 3 and 4, respectively. Example results obtained using the system for generating surface discharges are shown in Figures 5 and 6. In the presented Figures, along the ordinate, the energy value in keV is given, along the abscissa, the elapsed time is given, and the color bar is associated with the number of counts, which relates to the overall number (quantity) of energy pick of a particular energy value, registered by the device. The high-energy signals were recognized as locally increased density of counts, which were assumed to be caused by local breakdowns, thus PD occurred in the medium. Areas with increased high-energy radiation are marked with arrows. Additionally, some of the presented images do not show any significant changes in radiation, which is also marked on the respective images.

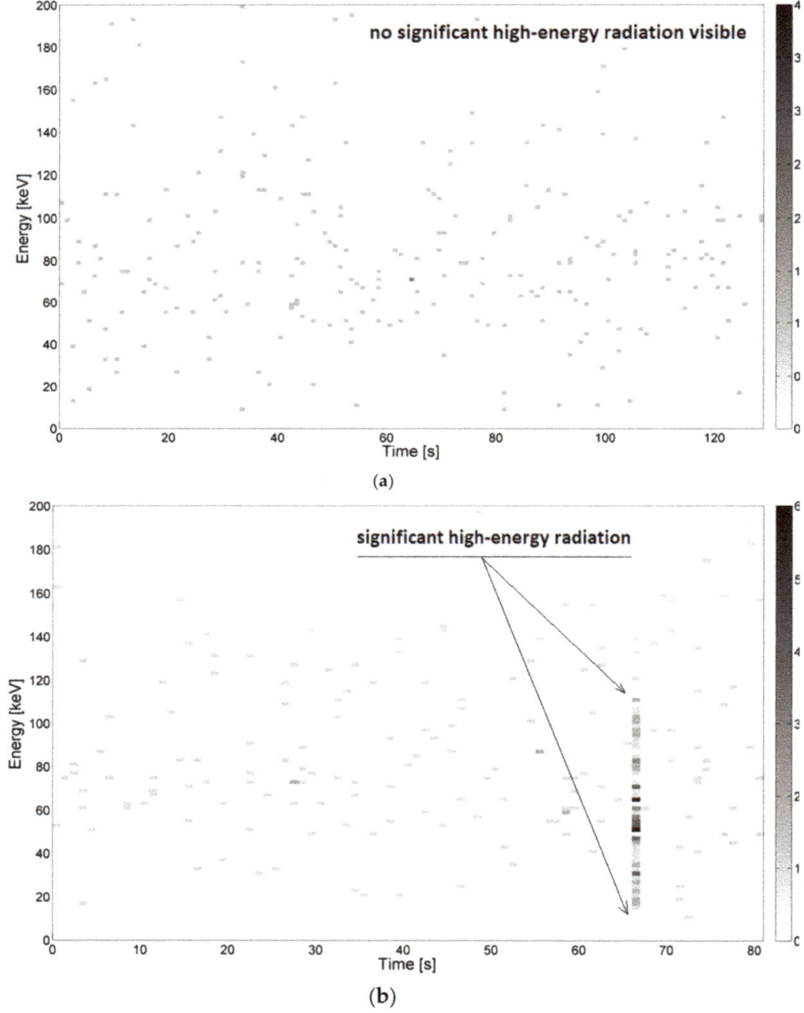

Figure 3. *Cont.*

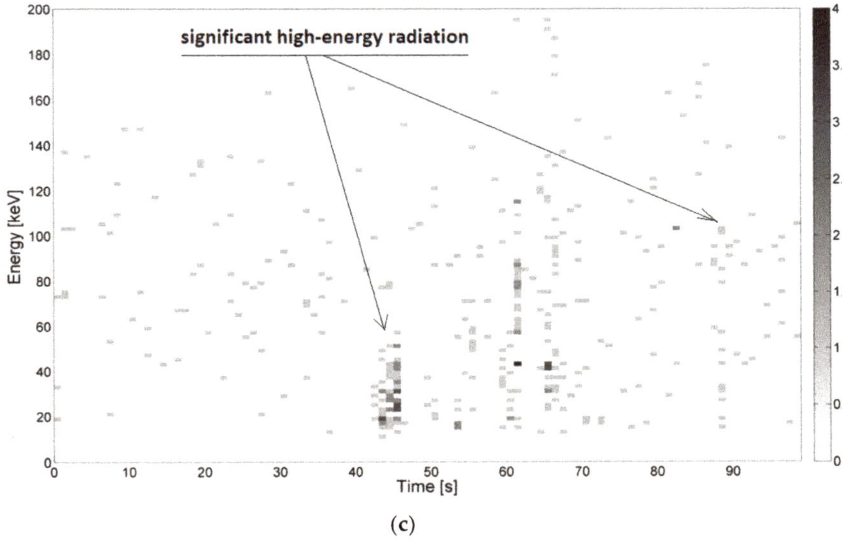

(c)

Figure 3. Example energy spectra of partial discharges (PD) in the needle–needle system for the following distances between the electrodes, (**a**) 100 mm, breakdown after t = 51 s (**b**) 180 mm, breakdown after t = 68 s, and (**c**) 200 mm, breakdown after t = 88 s. Supply voltage increased from 0 kV to 42 kV, 68 kV, and 72 kV.

Based on the achieved analysis, no significant differences in the value of recorded energy, measured in keV, and in the quantity (count no) of generated ionization events were observed. The electromagnetic field distribution between virtual electrodes might constitute a significant factor in this case.

Despite a high sensitivity of the applied high-energy radiation detector, no significant changes in the energy spectrum were noticed for the sphere–sphere system, whereas for the needle–needle system and the system for generating surface discharges noticeable changes were observed.

In Figure 3a–c, energy spectra measured during generation of PD for the needle–needle system mounted at the distances of 100 mm, 180 mm, and 200 mm are presented, respectively. For lower voltage values (see Figure 3a) no significant radiation was registered. From Figure 3b, it is recognized, that right before the breakdown, which occurred after about 68 s, a sudden change of the counted number of ionizing events was noticed, accompanying an increase in energy value up to about 70 keV. In Figure 3c, locally increased concentrations of ionizing events were also observed. The increased energy values were recorded for measurement times 43–48 s, 55–57 s, 60–70 s, and 89–91 s, which correspond to the values of voltage equal to about 40 kV, 45 kV, 50–55 kV, and 72 kV respectively. In other time periods, no significant radiation of energy recorded was noticed.

In Figure 4a–c, energy spectra measured during generation of PD for the sphere–sphere system mounted at the distances of 100, 180, and 200 mm are presented, respectively. As mentioned, for all investigated distances no significant changes in the energy spectrum and no locally increased densities of high-energy events were noticed.

In Figures 5 and 6, energy spectra measured during the generation of surface PD at the distance between electrodes being 3 mm (Figure 5) and 10 mm (Figure 6) are presented respectively. For both presented cases, the detector was localized at a distance of 30 mm to the PD source.

From Figure 5, it is recognized at time t, in the range 43–48 s, the values of energy are reaching from 10 keV to 30 keV. These signals were emitted by supply voltage ranging from 3.3 kV to about 7 kV. The most significant emission (up to 9 counts per ionizing event) was registered when the radiation detector was moved further from the PD source, at a distance equal to 30 mm. In Figure 6, it was observed that after time $t = 38$ s, while supplying the testbed with voltage values from 30 kV to 42 kV, the registered energy signals reached values from 10 keV to 40 keV. In this case, ionizing radiation was not recognized during the generation of discharges at lower voltages.

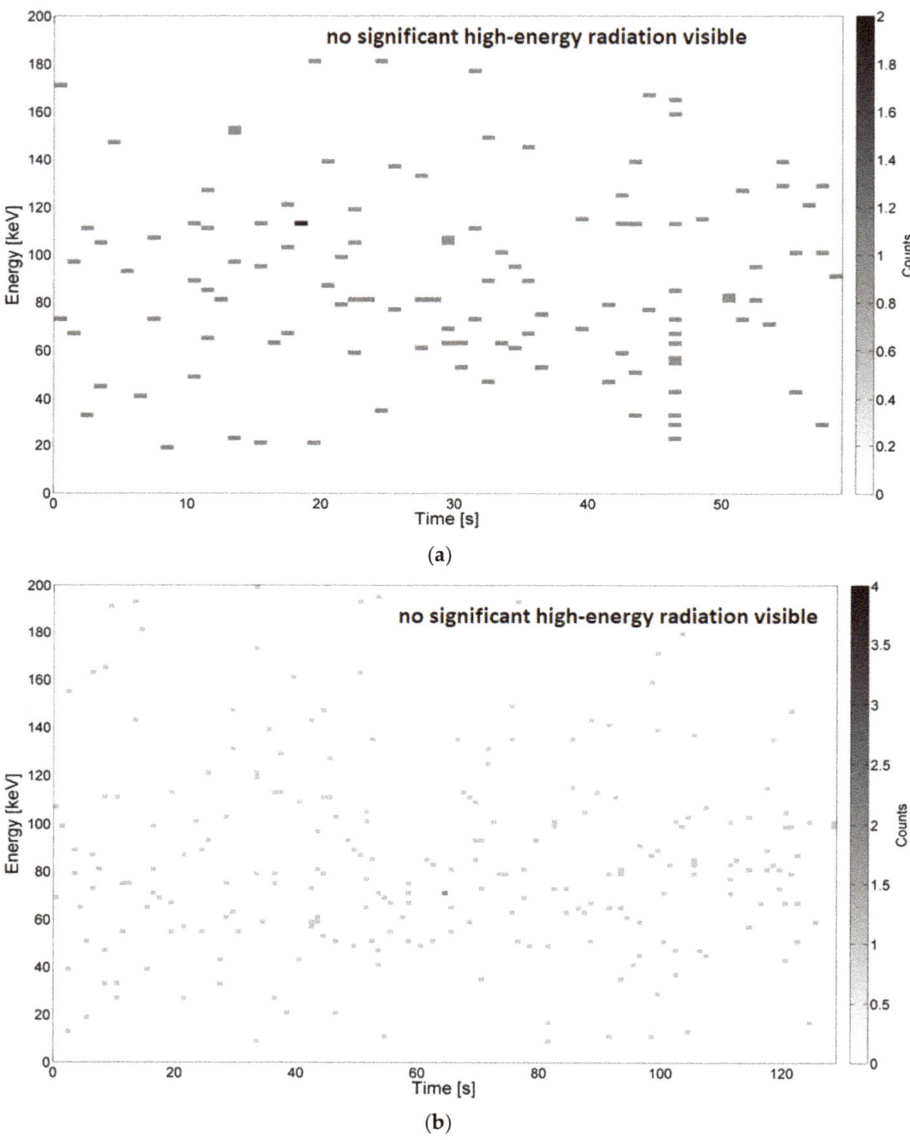

Figure 4. *Cont.*

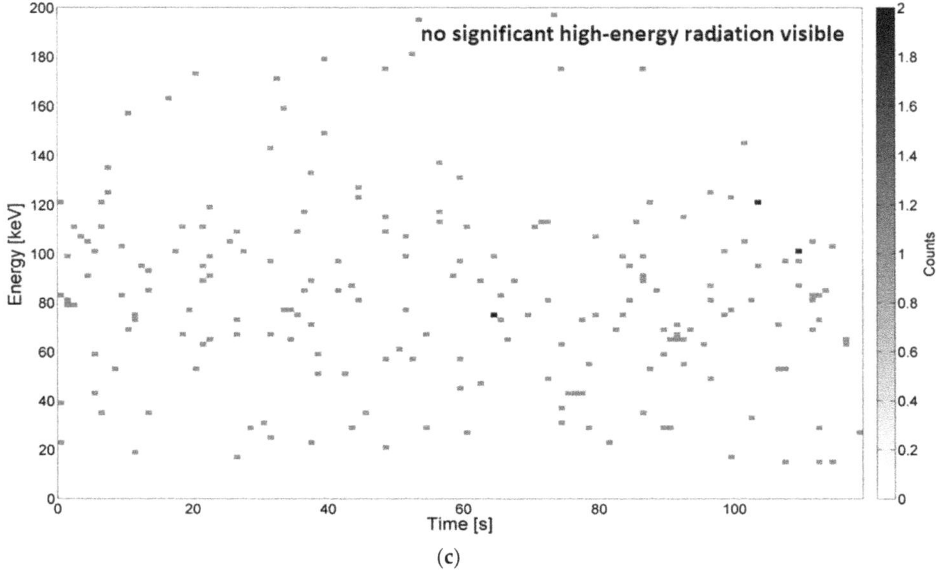

Figure 4. Example energy spectra of PD in the sphere–sphere system (50 mm) for the following distances between electrodes, (**a**) 100 mm, breakdown after t = 104 s, (**b**) 180 mm, breakdown after t = 102 s, and (**c**) 200 mm, breakdown after t = 105 s. Supply voltage increased from 0 kV to 64 kV, 83 kV, and 88 kV.

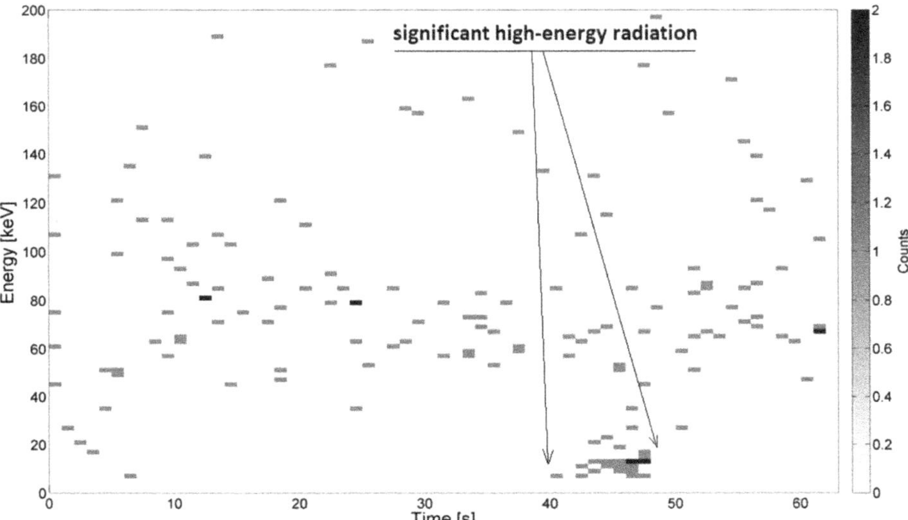

Figure 5. Example energy spectra of surface PD for the distance between the electrodes equal to 3 mm, supply voltage increased from 0 to 36 kV, breakdown after t = 49 s. The distance between the detector and the PD source was 30 mm.

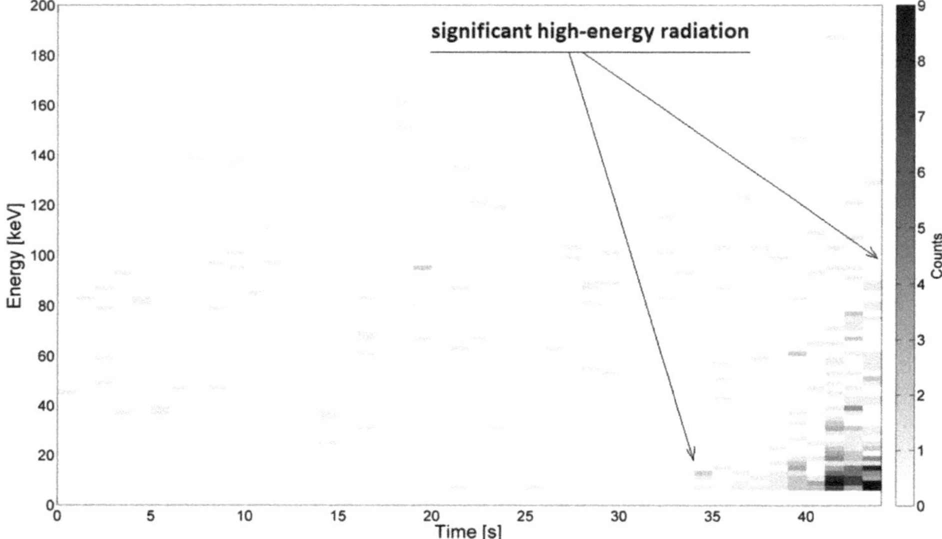

Figure 6. Example energy spectra of surface PD for the distance between the electrodes equal to 10 mm, supply voltage increased from 0 to 38 kV, breakdown after $t = 45$ s. The distance between the detector and the PD source was 30 mm.

5. Conclusions and Summary

The results obtained during research studies of ionizing radiation generated by PD in air were presented in this paper. The main aim was to investigate if and how much high-energy radiation is emitted by PD. Three PD generation systems were built. In a number of experiments, the influence of the distance between the high-voltage and ground electrodes, the distance between the radiation detector and PD source, and the supply voltage value on the registered radiation were estimated. Based on the achieved results the following statements were made:

- For the needle–needle spark gap system, local concentrations of the counted ionizing radiation events were noticed during the increase of supply voltage level. This was due to local breakdowns (PD) of the system tested. The mentioned indicates that the high-energy component exists and should be included as one of the phenomena accompanied by PD. This information is crucial e.g., for one who aims to investigate the balance of PD generation energy. Furthermore, the results presented suggest that the highest counted number and the highest energy was recorded right after the system breakdown;
- The energy spectra obtained during the generation of surface discharges also confirm the existence and measurability of the ionizing radiation;
- However, for the sphere–sphere spark gap system, such a characteristic spectrum was not obtained. One possible reason for it may be in the different electromagnetic field distribution between electrodes of various shapes (needle, sphere, and plate).

In general, we can state that an increase in the supply voltage did not have had any effect on the possibilities of recording ionizing radiation events. Furthermore, changes in the distance between the detector and the PD source, and changes in the distance between the electrodes, have not shown any unambiguous correlation or impact on the registered high-energy radiation. Nevertheless, it may be assumed that by supplying the setup with higher voltage values and by increasing the distance between the electrodes, the higher energy may be recorded. It may also be assumed that the distance between the detector and PD source has an effect on the number of recorded scintillation events and their energy value.

In future studies, research on the elemental composition of the electrodes, which may affect the achieved results, may be carried out. Moreover, the impact of humidity variations and other factors affecting the nature of the PD phenomenon may be interesting for investigation.

An important effect of detecting ionizing radiation and measuring its energy may be the correlation between the intensity of this radiation and the rate of degradation of the insulation of power lines. So far, the destructive effect of temperature and mechanical damage during discharges has been observed. The widening of known factors influencing the destruction of insulation is an important step towards the development of the discipline of electrical engineering and the reduction of operating costs of power equipment and power lines.

Another important element of the presented research is the possibility of using scintillation detectors in places where E-M radiation may come from another source. Determination of the spectra from electric discharges will allow for a calm assessment of the radiation source in places such as nuclear power plants, etc.

Author Contributions: Conceptualization, Ł.N., M.K. (Michał Kozioł); methodology, Ł.N., M.K. (Michał Kunicki); formal analysis, Ł.N., D.W., and M.K. (Michał Kunicki); validation, Ł.N., M.K. (Michał Kozioł), and M.K. (Michał Kunicki); resources, D.W., M.K. (Michał Kozioł); data curation, Ł.N., D.W.; writing—original draft preparation, Ł.N., M.K. (Michał Kozioł), and D.W.; writing—review and editing, Ł.N., M.K. (Michał Kunicki) and D.W.; project administration, Ł.N.

Funding: The work was co-financed from funds of the National Science Centre Poland (NCS) as part of the PRELUDIUM research project No. 2014/15/N/ST8/03680.

Conflicts of Interest: The authors declare no conflict of interest. The funders had no role in the design of the study; in the collection, analyses, or interpretation of data; in the writing of the manuscript, or in the decision to publish the results.

References

1. *IEEE Recommended Practice for Partial Discharges Measurement in Liquid-Filled Power Transformers and Shunt Reactors*; C57.113; IEEE Std.: Piscataway, NJ, USA, 2010.
2. Liang, R.; Wu, S.; Chi, P.; Peng, N.; Li, Y. Optimal placement of UHF sensors for accurate localization of Partial Discharges source in GIS. *Energies* **2019**, *12*, 1173. [CrossRef]
3. Kunicki, M. Variability of the UHF Signals Generated by Partial Discharges in Mineral Oil. *Sensors* **2019**, *19*, 1392. [CrossRef] [PubMed]
4. Siegel, M.; Coenen, S.; Beltle, M.; Tenbohlen, S.; Weber, M.; Fehlmann, P.; Hoek, S.M.; Kempf, U.; Schwarz, R.; Linn, T. Calibration Proposal for UHF Partial Discharges Measurements at Power Transformers. *Energies* **2019**, *12*, 3058. [CrossRef]
5. Loubani, A.; Harid, N.; Griffiths, H.; Barkat, B. Simulation of Partial Discharges Induced EM Waves Using FDTD Method—A Parametric Study. *Energies* **2019**, *12*, 3364. [CrossRef]
6. Dincer, S.; Duzkaya, H.; Tezcan, S.S.; Dincer, M.S. Analysis of Insulation and Environmental Properties of Decomposition Products in SF6-N2 Mixtures in the Presence of H2O. In Proceedings of the 2019 IEEE International Conference on Environment and Electrical Engineering and 2019 IEEE Industrial and Commercial Power Systems Europe (EEEIC/I&CPS Europe), Genova, Italy, 10–14 June 2019; pp. 1–6.
7. Nikjoo, R.; Mahmoudi, N. Relation of Winding Resistance Measurement and Dissolved Gas Analysis for Power Transformers. In Proceedings of the 2019 IEEE Milan PowerTech, Milan, Italy, 23–27 June 2019; pp. 1–6.
8. Wotzka, D.; Koziol, M.; Nagi, L.; Urbaniec, I. Experimental analysis of acoustic emission signals emitted by surface Partial Discharges in various dielectric liquids. In Proceedings of the 2018 IEEE 2nd International Conference on Dielectrics (ICD), Budapest, Hungary, 1–5 July 2018; pp. 1–5.
9. Ma, G.; Zhou, H.; Zhang, M.; Li, C.; Yin, Y.; Wu, Y. A High Sensitivity Optical Fiber Sensor for GIS Partial Discharges Detection. *IEEE Sens. J.* **2019**, *19*, 9235–9243. [CrossRef]

10. Yao, P.; Zheng, H.; Yao, X.S.; Ding, Z. A method of monitoring Partial Discharges in switchgear based on ozone concentration. *IEEE Trans. Plasma Sci.* **2019**, *47*, 654–660. [CrossRef]

11. Kozioł, M.; Boczar, T.; Nagi, L. Identification of electrical discharges forms, generated in insulating oil, using the optical spectrophotometry method. *IET Sci. Meas. Technol.* **2019**, *13*, 416–425. [CrossRef]

12. Koziol, M.; Nagi, L. Analysis of Optical Radiation Spectra Emitted by Electrical Discharges, Generated by Different Configuration Types of High Voltage Electrodes. In Proceedings of the 2018 IEEE 2nd International Conference on Dielectrics (ICD), Budapest, Hungary, 1–5 July 2018; pp. 1–4.

13. Nagi, Ł.; Kozioł, M.; Wotzka, D. Analysis of the spectrum of electromagnetic radiation generated by electrical discharges. *IET Sci. Meas. Technol.* **2019**, *13*, 812–817. [CrossRef]

14. Joseph, J.; Rajeevan, A.E.; Sindhu, T.K. Modelling and experimental validation of EM wave propagation due to PDs in XLPE cables. *IET Sci. Meas. Technol.* **2019**, *13*, 692–699. [CrossRef]

15. Tehlar, D.; Riechert, U.; Behrmann, G.; Schraudolph, M.; Herrmann, L.G.; Pancheshnyi, S. Pulsed X-ray induced Partial Discharges diagnostics for routine testing of solid GIS insulators. *IEEE Trans. Dielectr. Electr. Insul.* **2013**, *20*, 2173–2178. [CrossRef]

16. Shin, J.; Kim, S.; Kim, E.; Jo, H.; Sample, A.V. Partial Discharges Induction with X-rays to Detect Void Defects in Solid Insulating Materials. In Proceedings of the 2018 Condition Monitoring and Diagnosis (CMD), Perth, Australia, 23–26 September 2018; IEEE: Piscataway, NJ, USA, 2018; pp. 2–4.

17. Galli, G.; Hamrita, H.; Jammes, C.; Kirkpatrick, M.J.; Odic, E.; Dessante, P.; Molinie, P.; Cantonnet, B.; Nappé, J.C. Characterization and Localization of Partial-Discharge-Induced Pulses in Fission Chambers Designed for Sodium-Cooled Fast Reactors. *IEEE Trans. Nucl. Sci.* **2018**, *65*, 2412–2420. [CrossRef]

18. Cherepy, N.J.; Seeley, Z.M.; Payne, S.A.; Swanberg, E.L.; Beck, P.R.; Schneberk, D.J.; Stone, G.; Wihl, B.M.; Fisher, S.E.; Hunter, S.L.; et al. Transparent Ceramic Scintillators for Gamma Spectroscopy and Imaging. In Proceedings of the 2017 IEEE Nuclear Science Symposium and Medical Imaging Conference (NSS/MIC), Atlanta, GA, USA, 21–28 October 2017; IEEE: Piscataway, NJ, USA, 2018; pp. 2–3.

19. Cherepy, N.J.; Martinez, H.P.; Sanner, R.D.; Beck, P.R.; Drury, O.B.; Swanberg, E.L.; Payne, S.A.; Hurlbut, C.R.; Morris, B. New Plastic Scintillators for Gamma Spectroscopy, Neutron Detection and Imaging. In Proceedings of the 2017 IEEE Nuclear Science Symposium and Medical Imaging Conference (NSS/MIC), Atlanta, GA, USA, 21–28 October 2017; IEEE: Piscataway, NJ, USA, 2018; pp. 98–100.

20. Tisseur, D.; Estre, N.; Tamagno, L.; Eleon, C.; Eck, D.; Payan, E.; Cherepy, N. Performance evaluation of several well-known and new scintillators for MeV X-ray imaging. In Proceedings of the 2018 IEEE Nuclear Science Symposium and Medical Imaging Conference Proceedings (NSS/MIC), Sydney, Australia, 10–17 November 2018; IEEE: Piscataway, NJ, USA, 2019; pp. 1–3.

21. Sibczyński, P.; Moszyńki, M.; Szczęśniak, T.; Czarnacki, W.; Świderski, Ł; Schotanus, P. Properties of NaI(Tl) scintillator at liquid nitrogen temperature. In Proceedings of the 2011 IEEE Nuclear Science Symposium Conference Record, Valencia, Spain, 23–29 October 2011; IEEE: Piscataway, NJ, USA, 2012; pp. 1616–1620.

22. Liang, F.; Hoy, L. Comparison of NaI coupled to photomultiplier tube and silicon photomultiplier. In Proceedings of the 2018 IEEE Nuclear Science Symposium and Medical Imaging Conference Proceedings (NSS/MIC), Sydney, Australia, 10–17 November 2018.

23. Nagi, Ł.; Zmarzly, D.; Boczar, T.; Frącz, P. Detection of High-Energy Ionizing Radiation Generated by Electrical Discharges in Oil. *IEEE Trans. Dielectr. Electr. Insul.* **2016**, *23*, 2036–2041. [CrossRef]

Article

A Bayesian Approach for Remote Depth Estimation of Buried Low-Level Radioactive Waste with a NaI(Tl) Detector

Jinhwan Kim, Kyung Taek Lim, Kyeongjin Park and Gyuseong Cho *

Department of Nuclear & Quantum Engineering, Korea Advanced Institute of Science and Technology, 291, Daehak-ro, Yuseong-gu, Daejeon 34141, Korea; kjhwan0205@kaist.ac.kr (J.K.); kl2548@kaist.ac.kr (K.T.L.); myesens@kaist.ac.kr (K.P.)
* Correspondence: gscho@kaist.ac.kr

Received: 8 November 2019; Accepted: 2 December 2019; Published: 5 December 2019

Abstract: This study reports on the implementation of Bayesian inference to improve the estimation of remote-depth profiling for low-level radioactive contaminants with a low-resolution NaI(Tl) detector. In particular, we demonstrate that this approach offers results that are more reliable because it provides a mean value with a 95% credible interval by determining the probability distributions of the burial depth and activity of a radioisotope in a single measurement. To evaluate the proposed method, the simulation was compared with experimental measurements. The simulation showed that the proposed method was able to detect the depth of a Cs-137 point source buried below 60 cm in sand, with a 95% credible interval. The experiment also showed that the maximum detectable depths for weakly active 0.94-μCi Cs-137 and 0.69-μCi Co-60 sources buried in sand was 21 cm, providing an improved performance compared to existing methods. In addition, the maximum detectable depths hardly degraded, even with a reduced acquisition time of less than 60 s or with gain-shift effects; therefore, the proposed method is appropriate for the accurate and rapid non-intrusive localization of buried low-level radioactive contaminants during in situ measurement.

Keywords: remote-depth profiling; gamma spectral analysis; Bayesian inference; uncertainty estimation; radioactive nuclear waste; radiological characterization; nuclear decommissioning; radiation detection; low-resolution detector

---

1. Introduction

During the life cycle of nuclear facilities, a significant amount of radioactive waste is generated, resulting in large-scale land and building contamination. Characterization of these wastes is critical for decommissioning those contaminated sites because it can provide essential information related to design specifications and project planning required for environmental restoration [1–3]. In particular, acquiring knowledge of the depth profiling of radioactive contaminants is critical for choosing the most cost-effective decommissioning strategy because removing surface contamination at varying depths can significantly reduce the total disposal volume [4]. However, depth profiling remains challenging because porous materials, such as the soil and concrete that entrain the contaminants, also act as shielding materials. In fact, examples of wastes commonly encountered during the decommissioning of nuclear facilities include wastes buried inside such porous materials [5–7].

Traditional destructive methods, such as logging and core sampling, have been used for depth estimation; however, they are expensive and time-consuming [7,8]. Thus, various non-destructive techniques have been developed for remote-depth profiling including the relative attenuation method [9,10], principal component analysis (PCA) [11–13], and the approximate three-dimensional linear-attenuation method [14–16]. The relative attenuation method takes the relative difference in the attenuations of two primary peaks (i.e., the 32-keV X-ray and 662-keV gamma-ray peaks of Cs-137) in a measured spectrum to find the depth profile. However, the use of X-rays limits not only the maximum detectable depth to less than 2 cm due to their high attenuation, but also the number of specific radioactive sources. On the other hand, the PCA method extracts two principal component coefficients related to the depth of the buried radioactive source from a set of previously measured spectra with different burial depths. The synthetic angle is then defined based on the extracted coefficients to estimate the source depth. However, this method cannot effectively estimate the depth of a source buried more than 5 cm beneath the surface. Finally, the approximate three-dimensional linear attenuation method employs the well-known linear attenuation model [17] in 3D coordinates combined with information obtained from multiple measurements of gamma ray intensities on the surface of the material in which the radioactive contaminant is buried. This approach shows a clear improvement over earlier methods in terms of the maximum detectable depth up to 12 cm in sand for a 8.89-µCi Cs-137 source. Nonetheless, the aforementioned methods provide point estimates for the penetration depth of radioactive contaminants. That is, they ignore the uncertainty invariably associated with statistical fluctuations arising from physical processes that can only be determined by performing tedious repetitive measurements. In addition, the maximum detectable depth of existing methods is insufficient to detect significant amounts of internal contamination buried deep within a substance [5]. Likewise, gain-shift effects can degrade the performance of the existing methods because almost all detector-based systems are sensitive to changes in ambient temperature.

Therefore, we propose an advanced remote-depth estimation method for measuring buried radioactive contamination using Bayesian inference. Both simulation and experimental testing were conducted using a low-resolution NaI(Tl) detector to evaluate the performance of the proposed method under many possible scenarios. In addition, this work also emphasizes the influence of data-acquisition time and gain shifts upon the depth-estimation process. Lastly, we evaluated the sensitivity of the proposed model in terms of the prior distributions.

2. Materials and Methods

2.1. Bayesian Inference

In statistical inference, there are two different approaches to probability interpretations, namely frequentist inference and Bayesian inference [18]. The frequentist inference is based on the idea that probability is equal to the expected frequency of occurrence over a long period. In this case, the frequentist assigns unknown parameters to fixed values. Thus, the frequentist inference does not allow probability statements about the parameters of a statistical process. For instance, the fact that a 95% confidence interval for the normal mean value is within a certain range does not mean that 95% of this confidence interval contains the true value [18]. Instead, what it means is that there is a 95% chance that, when computing confidence intervals repeatedly many times, the true mean would lie in the 95% confidence interval. By contrast, Bayesian inference is based on the idea that probability represents the degree of belief in an event. That is, this concept allows us to treat the parameters as random variables. In this sense, a Bayesian statistician would say that there is a 95% probability that the true value will fall within the 95% credible interval of the given data range.

The objective of Bayesian inference is to determine the posterior distribution $p(\theta|y)$ of random variables θ given prior distributions $p(\theta)$ and likelihood function $p(y|\theta)$. This consideration is well-represented in Bayes' theorem [19]:

$$p(\theta|y) = \frac{p(y|\theta)\,p(\theta)}{p(y)},\tag{1}$$

where $p(\theta)$ is the probability in our belief of θ without observation of the data y; $p(y)$ is an evidence or normalization constant, from which the probability of the data is determined by integrating all possible values of θ ; and $p(y|\theta)$ is the probability of observing y generated by a model with θ. In fact, $p(\theta|y)$ is the refined probability of our belief in θ once y has been taken into account.

The difficulty in applying Bayesian inference to many practical applications arises when the intractable high-dimensional integrals of the evidence must be computed. However, recent advances in computation and in marginal estimation techniques using variational inference make it possible to solve such problems. The underlying idea behind variational inference is to convert the computation of the posterior distribution into an optimization problem. First, a parameterized family of distributions $q(\theta; v)$ (or equivalently, a variational distribution) is postulated. Then, we find the member of that particular family that minimizes the Kullback–Leibler (KL) divergence [19] of the exact posterior distribution. In fact, the KL divergence measures the closeness of the two distributions:

$$KL\big(q(\theta; v)\big\|p(\theta|y)\big) = \mathbb{E}_q\left[\log \frac{q(\theta; v)}{p(\theta|y)}\right]$$

$$= -\big(\mathbb{E}_q[\log p(\theta, y)] - \mathbb{E}_q[\log q(\theta; v)]\big) + \log p(y), \tag{2}$$

where $p(y)$ is intractable but has a constant value. Thus, minimizing the KL divergence is now equivalent to maximizing the evidence lower bound (ELBO):

$$\mathcal{L}(v) = \mathbb{E}_q[\log p(\theta, y)] - \mathbb{E}_q[\log q(\theta; v)]. \tag{3}$$

To simplify the inference further, a mean-field approximation can be assumed, where the parameters can be fully factorized into independent parts:

$$q(\theta_1, \ldots, \theta_n) = \prod_{i=1}^{N} q_i(\theta_i). \tag{4}$$

Despite efforts at simplifying via mean-field approximation in the variational inference, model-specific derivations and implementations are still required, making the process rather complex. Automatic differentiation variational inference (ADVI) [20] can solve this problem by offering an algorithm for automatic solutions associated with variational inference. ADVI begins by transforming $p(\theta, y)$ into the unconstrained real-valued random variables $p(\theta, \zeta)$. This transformation removes all original constraints on the latent variables, allowing us to consider the Gaussian distribution. Then, ADVI recasts the gradient of the variational objective function as an expectation over q In this way, one can make use of Monte Carlo methods to approximate $\nabla_\theta \log(\theta, y)$. Further re-parameterization of the gradient in terms of a standard Gaussian is performed for the purpose of achieving efficient computation of the Monte Carlo approximations. Finally, ADVI uses a stochastic gradient optimization approach for the variational distribution. We implemented ADVI in the Python language together with the probabilistic programming framework of PyMC3 [21]. This powerful library allowed us to easily specify a probability model and then perform variational inference.

2.2. Likelihood Function and Priors

In order to compute the posterior distribution of the depth of a buried radioactive source, we begin by defining a mathematical model that describes an observed spectrum in terms of the burial depth, activity, and relative shift magnitude of the spectrum (and hence, the shift). The spectrum measured from a collimated detector can be calculated as follows:

$$M_i = N_0 \delta \varepsilon = AP\delta \varepsilon = \frac{AP\delta}{4\pi h^2} e^{-\mu_A h} f(z, \eta i) + cB_i \text{ for } i = 1, \ldots, K, \tag{5}$$

where M_i is the measured spectrum (s^{-1}); i is the channel or energy index $(0 < i \le K)$; N_0 is the total intensity of gamma rays originally emitted from the source (s^{-1}); δ is the effective front area of the detector (cm^2); ε is the correction coefficient due to the attenuation in a material, gain-shift effects caused by temperature, and the inverse-square law; A is the activity in the radioisotope of interest per decay (μCi); P is the total gamma emission probability (i.e., 2 γs^{-1} Bq^{-1} for the 1173- and 1333-keV gamma rays of Co-60); μ_A is the linear attenuation coefficient of gamma rays in air (cm^{-1}); h is the detection height between the detector and the surface of a material (cm); z is the buried depth of a radioactive source $(0 \le z \le D$ cm) in a section of material from the front surface; η is the relative magnitude of the shift in the spectrum; B_i is the background spectrum in the measurement environment with K channels; c is its proportionality constant; and $f(z,\ \eta i)$ is the function for bilinear interpolation. To compute $f(z,\ \eta i)$ the $K \times D$ response matrix (or equivalently, the reference spectra), for a radioisotope should be determined by measuring the spectra at different depths ranging from 0 to D cm with certain intervals. It should be noted that all spectra obtained for the response matrix were normalized to the total count of the spectrum measured at a depth of 0 cm. Consequently, $f(z,\ \eta i)$ can be calculated as follows:

$$f(z,\ \eta i) = \frac{1}{(z_H - z_L)(i_H - i_L)}[z_H - z \quad z - z_L]\begin{bmatrix} f(z_L, i_L) \ f(z_L, i_H) \\ f(z_H, i_L) \ f(z_H, i_H) \end{bmatrix}\begin{bmatrix} i_H - i \\ i - i_L \end{bmatrix}. \tag{6}$$

As shown in Figure 1, $f(z,\ \eta i)$ is the interpolation point. In addition, $f(z_L, i_L)$, $f(z_H, i_L)$, $f(z_L, i_H)$, and $f(z_H, i_H)$ are the closest points to the $f(z,\ \eta i)$ among the known points from the $K \times D$ response matrix. In other words, the spectrum of 512 channels with the buried depth of the source at z cm and the shifted magnitude of η in the spectrum can be expressed as:

$$M = \frac{AP\delta}{4\pi h^2}[f(z,\ \eta),\ f(z,\ 2\eta),\dots,\ f(z,\ 512\eta)] + c\mathbf{B}. \tag{7}$$

Here, the value δ can be calculated from a laboratory experiment by placing a radioactive source on the surface of the material (at 0 cm in depth), and can be expressed as:

$$\delta = \frac{4\pi r^2 N}{APe^{-\mu_A r}}, \tag{8}$$

where N is the total net counts in the spectrum (s^{-1}) and r is the detection height (cm).

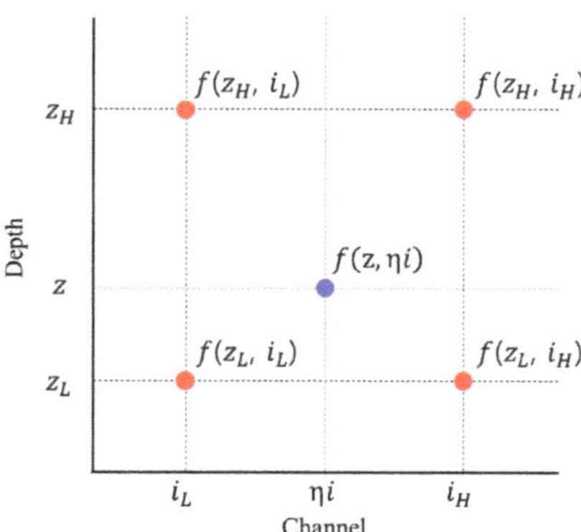

Figure 1. Example of bilinear interpolation for function f on target point $(z,\eta i)$ in a rectangular constellation formed by four grid points $(f(z_L, i_L),\ f(z_H, i_L),\ f(z_L, i_H),$ and $f(z_H, i_H))$.

As a matter of fact, our model prescribes the function $f(z, A, \eta, c)$. Nonetheless, the measured spectrum in practice experiences inevitable interference from the presence of irreducible uncertainties arising from physical processes, such as radioactive disintegration. Thus, the spectrum can be assumed to have a normal distribution with a zero mean and variance of σ^2:

$$M \sim f(z, A, \eta, c) + N\left(0, \sigma^2\right),\tag{9}$$

$$P(M|z, A, \eta, c) = N\left(f(z, A, \eta, c), \sigma^2\right).\tag{10}$$

When the initial information about the distributions of z, A, η, c, and σ^2 is available, it should be included as the priors. Table 1 shows the priors prescribed by the model where A, c, and σ^2 should be continuous and positive-only, and the gamma distribution can therefore be selected accordingly. However, one may wonder how the parameters were determined, that is, a shape parameter α and an inverse scale parameter β in the gamma distribution, because the parameters represent our degree of belief. In this work, we divided the measured spectrum by the acquisition time and analyzed them on a one-second basis, which was true in case of the background spectrum. In this regard, the most probable value of c would be one. Therefore, the determination of the parameters with gamma(1, 1), i.e., the mean and variance are equal to one, would be reasonable. In addition, the parameters can be considered as being equal to gamma(1, 1) since σ^2 is expected to yield a small value. Although the choice on the parameters in the gamma distribution for A could be somewhat ambiguous, it would still be appropriate to determine the parameters with gamma(1, 1); this was because our belief is that the level of the source activity would be low. On a contrary, the parameters z and η should be confined within certain ranges. In fact, the range for z was selected from 0 to D cm, which was the maximum burial depth of the radioactive contaminant to be considered; the value D was selected as 50 cm (or 60 cm) such that the proposed method would search extensively for the burial depths of the source ranging from 0 to 50 cm. As for the parameter η, the value was chosen from a range of 0.85 and 1.15 in consideration of the relative Cs-137 peak positions in the NaI(Tl) spectra depending on the temperature ranging from 0 to 50 °C such that the selected range would contain all possible values for the shift magnitude [22]. It is worth emphasizing that any values for the parameters (i.e., α and β) in which their multiplication would be equal to one or other choices that would yield a small mean value in the gamma distribution is possible. In addition, other distributions, such as a truncated normal distribution and triangular distribution, could be selected based on knowledge of the evaluators. Since the selection on prior distributions could influence posterior distributions, the sensitivity analysis on prior distributions was necessary to verify the robustness of our model, which is presented in Section 3.5.

Table 1. The priors prescribed in the model.

Variable	Prior
z	uniform$(0, D)$
A	gamma$(1, 1)$
η	uniform$(0.85, 1.15)$
c	gamma$(1, 1)$
σ^2	gamma$(1, 1)$

2.3. Monte Carlo Modeling and Simulation

In order to validate the proposed method, we performed Monte Carlo modeling and simulations using Monte Carlo N-Particle Transport Code, version 6 (MCNP6) [23]. A schematic of the MCNP6 model used for the simulation is reported in Figure 2. The simulation model consists of a NaI(Tl) detector located 6 cm away from the surface of a box that is filled with sand of density 1.7 g cm^{-3} [24]. The detector itself is surrounded by a hollow cylindrical lead shield lined with copper for minimizing the scattered gamma rays and X-ray fluorescence from lead. A pulse-height tally was used to simulate the gamma-ray spectra. In addition, an FT8 Gaussian-energy-broadening card was applied to mimic the physical spectra as realistically as possible. In fact, this feature simulates the peak-broadening effect arising from a physical radiation detector based on coefficients "a," "b," and "c." These coefficients allow the MCNP6 to recognize the continuous values of the full width at half maximum (FWHM) in the energy range of interest based on the non-linear function in Equation (11), where E is the incident gamma-ray energy. We employed parametric optimization using a genetic algorithm to find the optimal value of these coefficients [25]:

$$\text{FWHM} = \text{a} + \text{b}\sqrt{E + cE^2}. \tag{11}$$

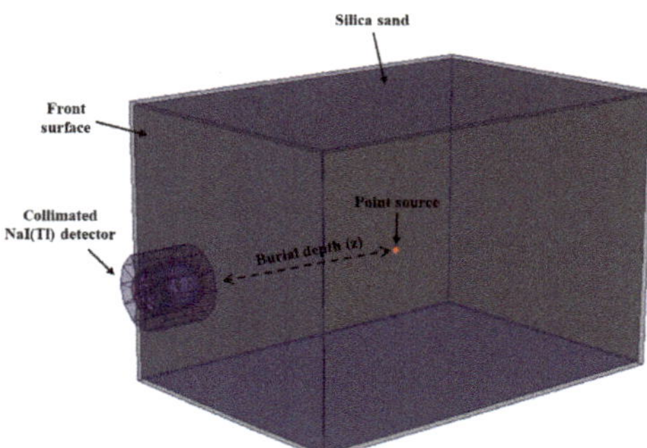

Figure 2. Schematic of the geometry defined for MCNP6 simulation. A collimated NaI(Tl) detector was placed 6 cm away from the front surface and a radioactive Cs-137 source was located inside silica sand.

The radioisotope Cs-137, which is one of the most predominant anthropogenic radioactive contaminants, arising as it does in fission fragments from spent nuclear fuel, was used and treated as a point source during the simulation. To obtain the response matrix, the point source was placed in sand at depths of 0, 1, 3, 5, 7, 10, 15, 20, 25, 30, 40, 50, and 60 cm. At each depth, a total of 3×10^9 particles were generated to achieve a sufficient counting statistic in the simulated spectra. For the test spectra, the point source was buried at a depth from 0 to 60 cm with 3-cm intervals. In this case, a total of 2×10^8 gamma particles were transported at each depth.

2.4. Experimental Setup

The experimental setup for the acquisition of gamma-ray spectra is shown in Figure 3a. The setup was composed of a sandbox filled with fine silica sand in which a radioactive source was buried and the two-inch NaI(Tl) detector was located 6 cm away from the box surface. The dimensions of the box were 50 cm × 40 cm × 40 cm (length × width × height), and it was constructed from acrylic sheets of thickness 0.3 cm. The use of acrylic and its thickness were chosen to allow almost all gamma rays to scatter exclusively within the sand matrix. In the experiment, a sealed Co-60 source was used in addition to Cs-137, because Co-60 is also found in nuclear environments resulting from the neutron activation of steel parts. The activity of the Cs-137 and Co-60 sources were 0.94 µCi and 0.69 µCi, respectively. The source was buried in a graduated box (50 cm × 3 cm × 3 cm) filled with sand, as shown in Figure 3b. Then, the box was inserted into the sandbox such that the exact distance of the source from the front of the sandbox would be achievable. The detector was placed inside a cylindrical lead collimator with a thickness of 2 cm, clad in a 0.5-cm-thick copper layer.

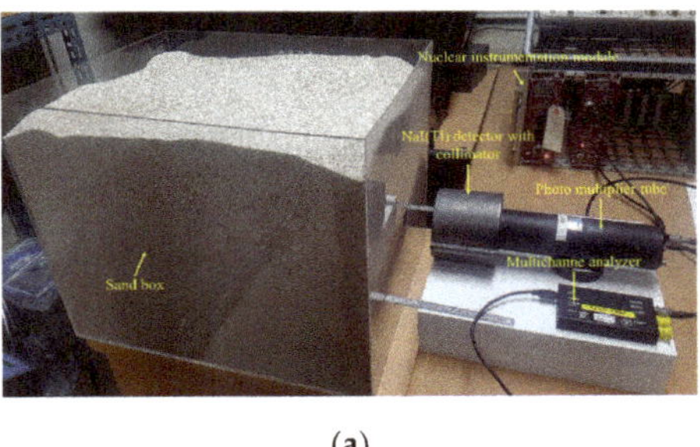

(a)   (b)

Figure 3. (**a**) Experimental setup for the gamma-ray measurement and (**b**) a graduated box for adjusting the burial depth of a radioactive source.

In the experiment, the spectra for the response matrix and the test set were measured at the same position of each radioactive source, as mentioned in Section 2.3, up to depths of 50 cm and 48 cm, respectively. The spectra for the response matrix were measured for 50 min to minimize statistical fluctuations in the spectra, while the test spectra were measured for 10 min. In addition, a background spectrum for the response matrix was also acquired under the same condition but without sources. Furthermore, the energy ranges of the spectra for Cs-137 and Co-60 were chosen from 300 to 1500 keV (i.e., 369 channels) and 700 to 1500 keV (i.e., 246 channels), respectively. During the experiment, energy calibration was not strictly performed because the proposed model was capable of compensating for gain-shift effects.

3. Results

3.1. Depth Estimation of the Buried Cs-137 Based on Simulated Spectra

Figure 4 shows the joint probability distributions of the depth and activity of Cs-137 sources for selected depths, where the red line represents the true values for the depths and activities. As expected, the joint distribution gradually spread out with a strong positive correlation with increasing source depth. This was mainly due to the increased number of gamma rays scattering in the sand matrix and the decreased number of detected photons, which consequently caused an increase in the statistical fluctuations in the spectra. Moreover, the positive correlation was clear because the activity was inversely related to the square of the depth (see Equation (5)). In this regard, the estimated values of the activity changed more sensitively when the source was deeply buried.

The estimated depth with a 95% credible interval for the Cs-137 source buried in sand over a range from 0 to 60 cm with 3-cm intervals is shown in Figure 5a. It shows that the true depth was well-approximated by the mean value of the estimated depth up to 45 cm. Beyond this depth, the estimated values tended to be underestimated. Nonetheless, all true depths fell within the 95% credible interval of the estimated depths until reaching a depth of 60 cm. In fact, the values for the source depths used to generate the response matrix rarely included those used to generate the test spectra. In other words, the interpolation method estimated the spectra that were not predetermined at certain depths well and therefore it was not necessary to take measurements for the response matrix at every single depth. As shown in Figure 5b, the estimated activities also closely agreed with the true activities, considering the 95% credible intervals. However, it is worth noting that the credible interval of the activities was more susceptible to fluctuation due to the inverse-square law.

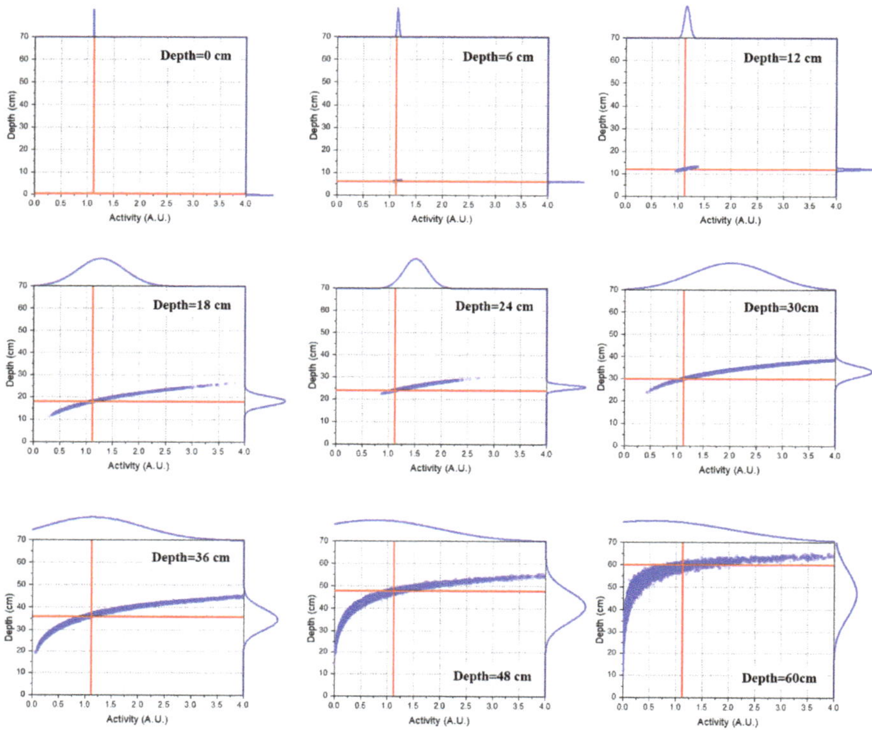

Figure 4. Joint probability distributions between the depth and activity of Cs-137 for selected depths simulated via MCNP6. The scatter dots in the central area depict the correlations between the estimated depth and activity with red lines representing true values, while the curves outside the plot area represent their corresponding densities.

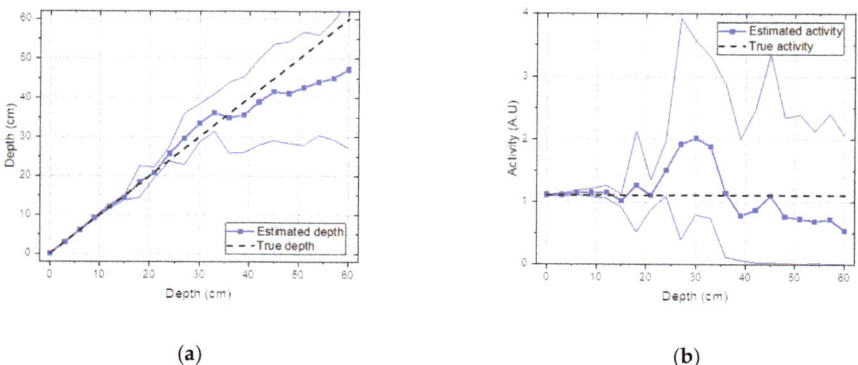

(a)   (b)

Figure 5. (a) Estimated depth and (b) activity with a 95% credible interval for spectra simulated with Cs-137 buried in sand from 0 to 48 cm with 3-cm intervals.

3.2. Depth Estimation of Buried Cs-137 and Co-60 Based on Experimental Spectra

The estimation of the joint probability distributions of the depth and activity of Cs-137 for selected depths is reported in Figure 6. From the plots, we can see a similar trend to the simulation results. In particular, this can be clearly seen in Figure 7a, in which the true depth was well-approximated by the estimated depth up to 21 cm. Beyond this, the value of the estimated depth began to be underestimated and to fluctuate. It is interesting to note that this depth was much lower than the value obtained using simulations. This was because the influence due to the attenuation and background became significant with increasing source depth, resulting in no remarkable feature differences in the acquired spectra, as illustrated in Figure 8a. In addition, the weak activity of the Cs-137 source used for the experiment could also be attributed to the obtained result. By contrast, noticeable feature differences were observed in the spectra at depths from 0 to 21 cm, as shown in Figure 8b. Based on the results, the maximum detectable depth of Cs-137 for the experimental setup was determined to be 21 cm.

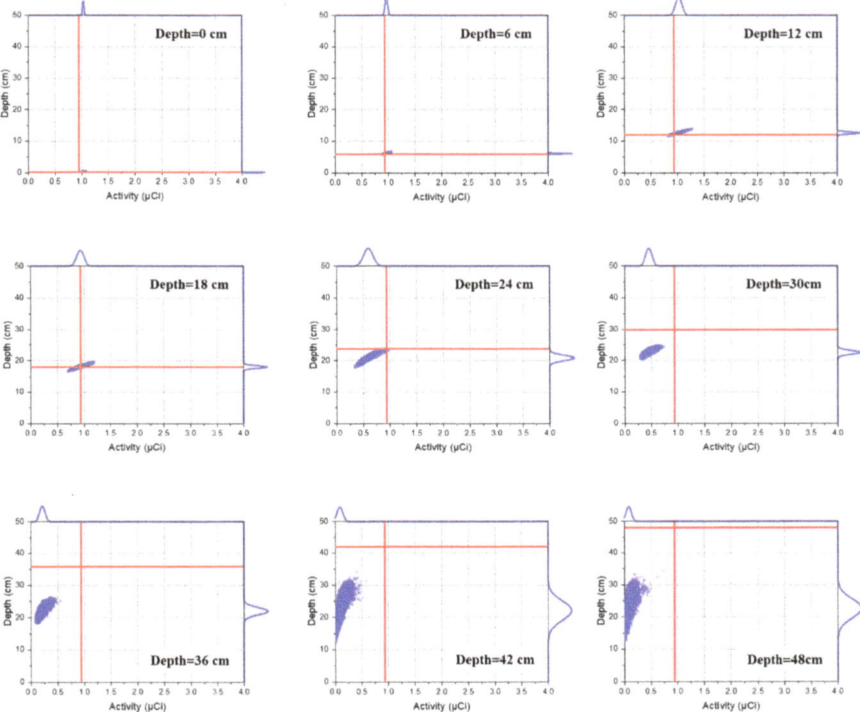

Figure 6. Joint probability distributions between the depth and activity of Cs-137 for selected depths, as measured experimentally. The scatter dots in the central area depict the correlations of estimated depth and activity, with red lines representing true values, while the curves outside the plot area represent their corresponding densities.

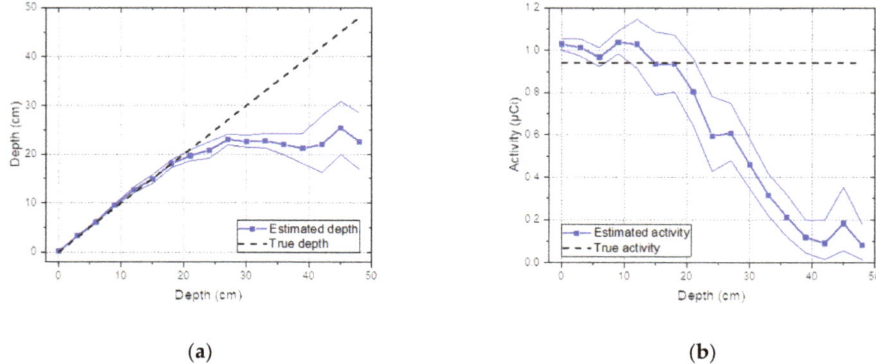

(a)   (b)

Figure 7. (**a**) Estimated depth and (**b**) activity with a 95% credible interval for spectra measured from Cs-137 buried in sand from 0 to 48 cm with 3-cm intervals.

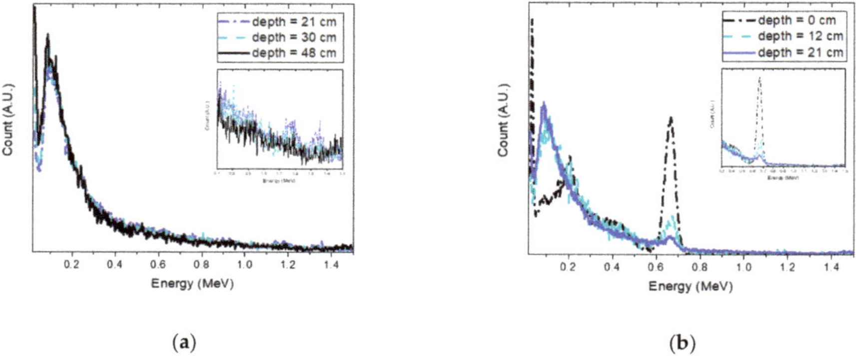

(a)   (b)

Figure 8. Normalized spectra for different depths of Cs-137 with a 600-s acquisition time: (**a**) the source was buried at depths of 21 cm, 30 cm, and 48 cm; and (**b**) the source was buried at depths of 0 cm, 12 cm, and 21 cm. The spectra were normalized to the total count across all energies for comparison purposes only. The inset shows the spectra within the energy range used for analysis.

The same experiment was also performed with the Co-60 source. Figure 9a shows the estimated depth with a 95% credible interval for the Co-60 source buried in sand from 0 to 48 cm with 3-cm intervals. As expected, the same trend was observed as in the Cs-137 case; the true depth was well-approximated by the mean value of the estimated depth up to 21 cm and the estimated depth gradually deviated from the true depth and fluctuated beyond the depth of 21 cm. The reason for this was the same as that mentioned earlier, namely differences in spectral features became negligible with increasing source depth, as shown in Figure 10.

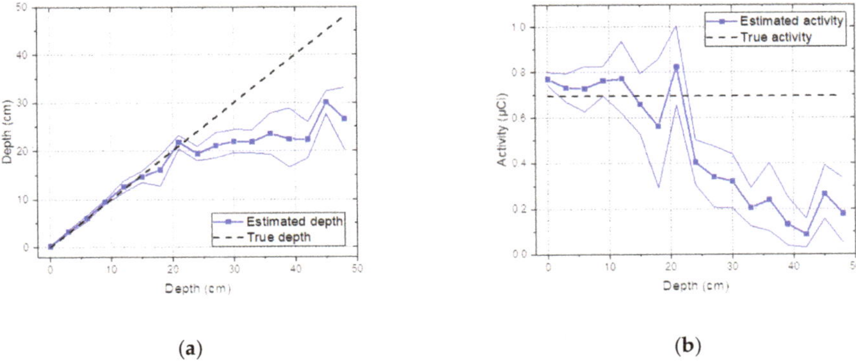

(a)   **(b)**

Figure 9. (**a**) Estimated depth and (**b**) activity with a 95% credible interval for spectra measured from Co-60 buried in sand from 0 to 48 cm with 3-cm intervals.

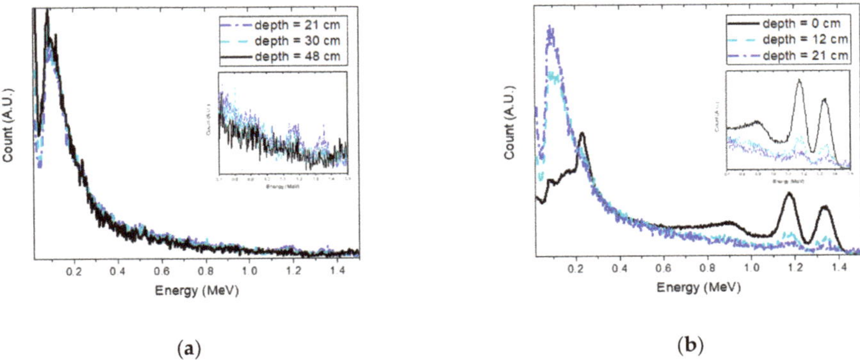

(a)   **(b)**

Figure 10. Normalized spectra for different depths of Co-60 with a 600-s acquisition time: (**a**) the source was buried at depths of 21 cm, 30 cm, and 48 cm; and (**b**) the source was buried at depths of 0 cm, 12 cm, and 21 cm. The spectra were normalized to the total count across all energies for comparison purposes only. The inset shows the spectra within the energy range used for analysis.

3.3. Effect of Acquisition Time

To perform an in-depth analysis of the sensitivity with respect to acquisition time, gamma-ray spectra with reduced acquisition times were analyzed at the same varying depths as those mentioned in Section 3.2. Figure 11a,b shows the estimated depths with 95% credible intervals analyzed for the spectra of Cs-137 with acquisition times of 10 s and 60 s, respectively. Despite the very short acquisition time, an identical trend was observed as in the spectra obtained over 600 s; the true depths were well-approximated for the mean values of the estimated depths up to 21 cm, which was the maximum detectable depth found in Section 3.1. Surprisingly, even at depths beyond the maximum detectable depth, the true depths seemed to yield a better approximation using the estimated depths with consideration of the 95% credible intervals. This may have been due to the statistical fluctuation in spectra caused by the reduced acquisition time, resulting in a more extensive search of the parameter space of the depth and activity. In fact, this is clearly seen in Figure 12b; the joint distributions gradually diverged to neighboring depths and activities with decreasing acquisition time. It is worth mentioning that 10-s acquisition times are extremely short for a typical in situ measurements, which can lead to a highly fluctuating spectrum, as reported in Figure 12a.

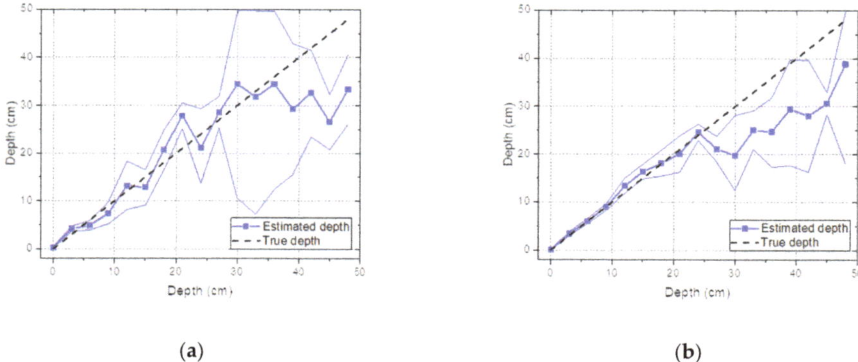

(**a**)   (**b**)

Figure 11. Estimated depth with a 95% credible interval analyzed for Cs-137 buried in sand over the range of 0 to 48 cm at 3-cm intervals from the experiment and the corresponding value of the true depth: (**a**) acquisition time of 10 s and (**b**) acquisition time of 60 s.

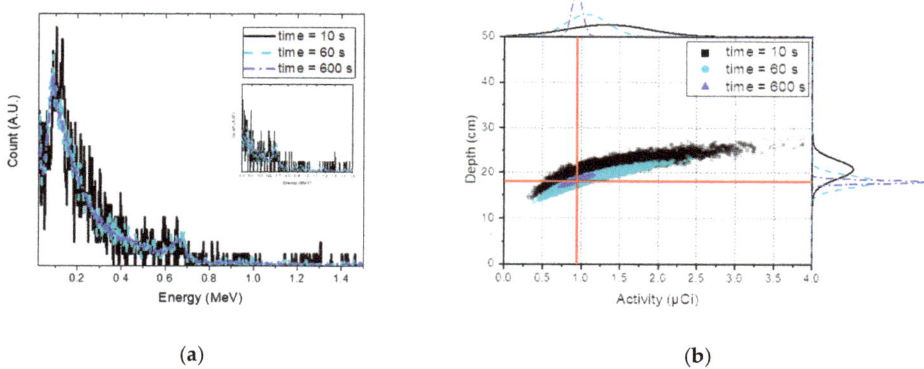

(**a**)   (**b**)

Figure 12. (**a**) Normalized spectra for Cs-137 buried at depths of 18 cm with acquisition times of 10 s (black solid line), 60 s (blue-sky dashed line), and 600 s (blue dash-dotted line). (**b**) Joint probability distributions between the depth and activity analyzed for the same spectra; the red line represents the true value of the depth and activity. The inset shows the spectra within the energy range used for analysis.

3.4. Effect of the Gain Shift

To demonstrate the ability to accommodate shifts in the spectra mainly due to temperature variations in the proposed model, spectra were measured with the Cs-137 source buried at a depth of 18 cm with the gain factor adjusted from 0.60 to 0.70 with 0.01 increments. Note that the test spectra were acquired with a gain factor of 0.65. In Figure 13a, we can see that the estimated depth fluctuated only slightly near the true depth for the spectra obtained with gain factors between 0.63 and 0.70. Furthermore, the true value of the depth fell within a 95% credible interval of the estimated depth within the investigated range. As expected, the trend of the change in the estimated shift increased with the increasing gain factor, as shown in Figure 13b. This was mainly because the model, combined with the bilinear interpolation method, scanned the shifted spectrum and searched for the probability distributions of the depth and shift that were most likely to have generated the spectrum via Bayesian inference. However, the estimated depth tended to increase and was thereby overestimated below the gain factor of 0.63. This was likely due to the slope connecting the Compton continuum and Compton valley in the spectrum becoming increasingly steep as the spectrum shifted in a negative direction, which is a typical phenomenon occurring when more gamma rays are scattered in a substance. Figure 14 shows an example of the spectra measured with gain factors of 0.63 and 0.70. It should be noted that these shifted spectra would be difficult to analyze without any recalibration.

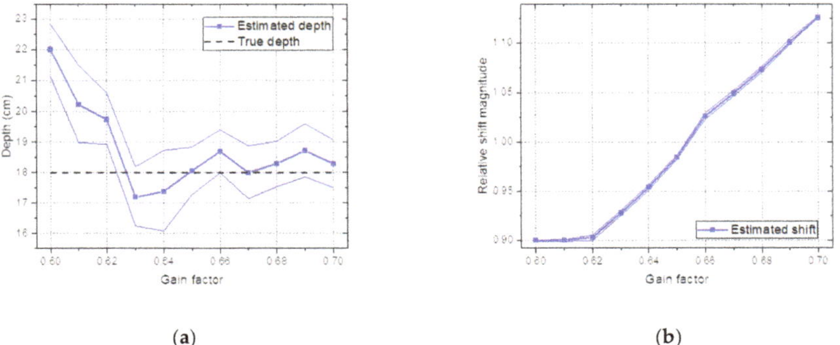

(a)   (b)

Figure 13. (a) Estimated depth and (b) estimated shift analyzed for the spectra of Cs-137obtained by changing the gain factor of the amplifier between 0.60 and 0.70 with 0.01 intervals.

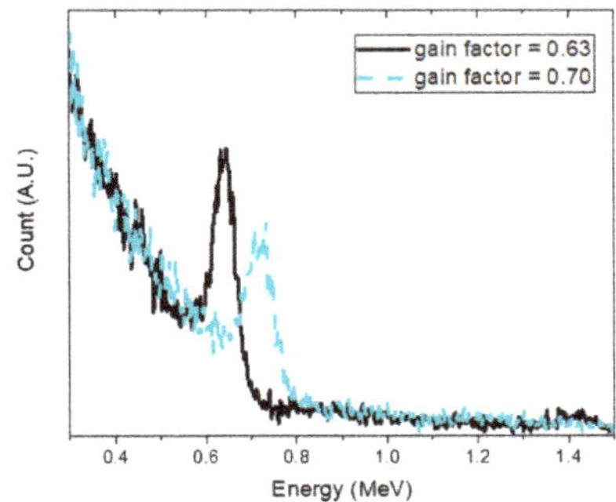

Figure 14. Example of the spectra of Cs-137 obtained with a gain factor of 0.63 (black solid line) and 0.70 (gray dashed line).

3.5. Sensitivity of Prior Distributions

To verify the robustness of our model with respect to the sensitivity to the prior distributions, experimental spectra for the Cs-137 source buried in sand from 0 to 48 cm with 3-cm intervals were analyzed using different prior distributions. Figure 15a shows the results using the following prior distributions: z followed a triangular distribution with parameters (0, 50, 4); A followed a gamma distribution with parameters (50, 0.2); η followed a uniform distribution with parameters (0.1, 1.1); c followed a gamma distribution with parameters (5, 5); and σ^2 followed a gamma distribution with parameters (0.1, 1). In other words, evaluators believed that the activity of Cs-137 source could be at a high level (i.e., the mean value was equal to 250 μCi) and buried deep (i.e., the mean value was equal to 18 cm). In addition, the observed spectrum could be shifted negatively. On the other hand, Figure 15b shows the results using the following prior distributions: A, η, and c all followed uniform distributions with parameters $(0, 10^{10})$; z followed a truncated normal distribution with parameters $(25, 10^{10}, 0, 50)$ supported by $z \in [0, 50]$; σ^2 followed a half-normal distribution with parameters $(0, 3)$ supported by $\sigma^2 \in [0, \infty]$. Such a wide range of distribution can be non-informative for the data on a small numerical scale. In other words, evaluators would have very limited information. Thus, we confirmed that the same trends were observed up until the maximum detectable depth (i.e., 21 cm), as seen from the Cs-137 case (see Figure 7a), regardless of their different beliefs; the estimated depth agreed well with the true depth up to 21 cm. It is worth emphasizing that the proposed model may not be completely free from the choice on the prior distributions. Nevertheless, evaluators with some knowledge of radiation measurements may have a minimal influence over the posterior distributions.

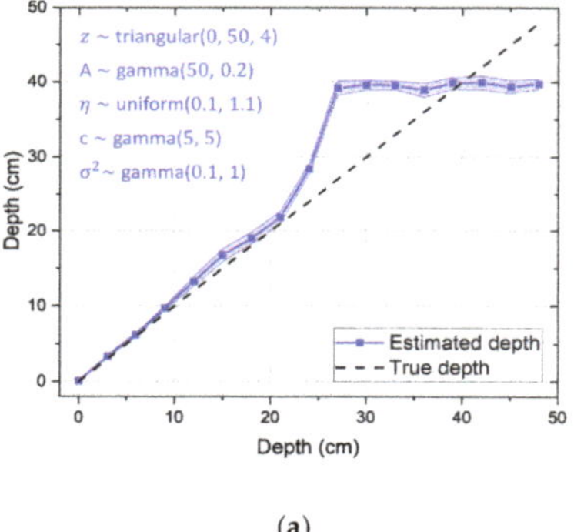

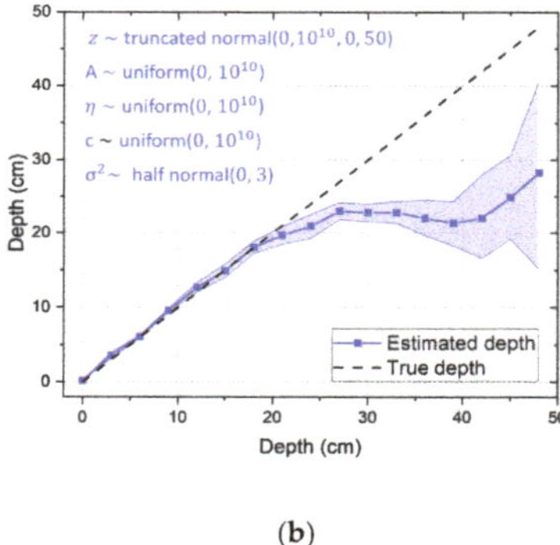

(a)

(b)

Figure 15. Estimated depths with a 95% credible interval for measured spectra for a Cs-137 source buried in sand from 0 to 48 cm with 3-cm intervals. These results were obtained with the prior distributions as reported within each figure. The letters marked in blue are the prior distributions that account for different beliefs.

4. Discussion

We presented an application of Bayesian inference to improve the estimation of remote depth profiling for low-level radioactive contaminants. From the simulation and experimental results, we confirmed that the proposed technique has significant advantages compared to existing methods for localized radioactive wastes. First, our approach allowed us to determine the probability distribution for parameters of interest, i.e., depth, activity, and shift, with improved reliability in a single measurement. From the measurement perspective, inherent uncertainty due to quantum physics is inevitable and must therefore be quantified. Thus, we should be able to provide measurement analysis in a statistical manner. However, previous depth estimation methods calculate uncertainty only via tedious repetitive measurements. Second, the proposed model yields a larger value for the maximum detectable depth. Recent studies have shown that the maximum detectable depths of 8.89-µCi Cs-137 and 0.24-µCi Co-60 buried in sand are 12 cm and 3 cm, respectively [14–16]. In fact, the activity intensity of the Cs-137 used in this work was about 10 times weaker than that of the Cs-137 used in those studies. Hence, the maximum detectable depth of 21 cm for both weakly active 0.94-µCi Cs-137 and 0.69-µCi Co-60 sources buried in sand was indeed a significant improvement in comparison to existing methods [9–16]. Third, the proposed technique provided a much faster and more accurate estimation of depths up to the maximum detectable depth (i.e., 21 cm), which was achieved within 60 s, even for sources with a weak activity. This is advantageous because radiological characterization for decommissioning involves scanning or measuring a wide range of sites, given that the activity of radioactive contaminants is not high enough in general. Fourth, the proposed technique was less susceptible to the gain shifts caused by temperature changes. One of the challenging issues for in situ measurement systems is that detectors are sensitive to changes in the ambient temperature, which can cause gain shifts. Therefore, regular quality control measurements become more critical to ensure a stable system operation. Thus, the present method automatically calibrated the degree of shift in the spectrum that would have been affected within the range of the prior distribution (see Table 1), which incorporated the magnitude of shifts in spectra due to temperature variation in the NaI(Tl) detectors [22]. In addition, the output of the shift could be used as a real-time indicator to demonstrate how stable the measurement systems are in operations on site. Finally, we confirmed that our model was not very sensitive to the choice of the prior distributions such that evaluators with some knowledge of radiation measurements would be able to obtain similar results. This feature is important for many practical applications because a Bayesian model is said to be non-robust and sensitive depending on the prior distributions.

However, this technique has some difficulties. One of the challenges in applying this technique lies in the establishment of a response matrix for materials in which a certain radioactive source is buried. To do this, the setup of materials affecting the attenuation of gamma rays should be carefully performed in order to mimic a real environment. Another difficulty is that the detector cannot simply be located at an optimal position in which the radioactive contaminants are buried; however, this can be resolved by placing the detector at the position in which the maximum intensity of the total count rate would likely be observed by taking a uniform scanning time. Lastly, most remote depth estimation methods assume that only a single radioisotope exists and that no other radioisotopes interfere with the measurement. In practice, such assumptions are sometimes not applicable. Furthermore, foreknowledge of radioisotopes present at a site is not available in some cases. Therefore, a better solution would be to integrate the likelihood function for quantitative analysis of radioisotopes, as proposed in Kim et al. [26] with the likelihood function used in this particular work. This solution will enable an accurate depth estimation for multiple radioactive sources without foreknowledge of the radioisotopes present.

5. Conclusions

In this work, we demonstrated an advanced method for remote depth estimation of localized radioactive contaminants using Bayesian inference. This approach, which is completely different from frequentist inference, allowed us to estimate the uncertainty of the depth and activity via a single measurement. The results of the simulation and experiment for Cs-137 and Co-60 sources buried in sand showed a significant improvement in the maximum detectable depth compared to those of existing methods. In addition, experimental results confirmed that the level of accuracy and the depth limit were preserved, even with a short acquisition time. Furthermore, the proposed technique was capable of accommodating for gain-shift effects caused by temperature variations, enabling a rapid non-intrusive localization of buried radioactive contaminants during in situ measurements as a consequence.

Author Contributions: Conceptualization, J.K. and G.C.; methodology, J.K.; software, J.K. and K.P.; validation, J.K. and K.T.L.; formal analysis, J.K.; investigation, J.K.; writing—original draft preparation, J.K.; writing—review and editing, G.C. and K.T.L.; visualization, J.K. and K.P.; supervision, G.C. and K.T.L.; funding acquisition, G.C.

Funding: This work was supported by the Center for Integrated Smart Sensors funded by the Ministry of Science and ICT as Global Frontier Project (CISS-2016M3A6A6929965) and NRF (National Research Foundation of Korea) Grant funded by the Korean Government(NRF-2018-Global Ph.D. Fellowship Program).

Conflicts of Interest: The authors declare no conflict of interest.

References

1. Characterization of Radioactively Contaminated Sites for Remediation Purposes. Available online: https://www-pub.iaea.org/MTCD/publications/PDF/te_1017_prn.pdf (accessed on 5 September 2019).
2. Radiological Characterisation for Decommissioning of Nuclear Installations. Available online: https://www.oecd-nea.org/rwm/docs/2013/rwm-wpdd2013-2.pdf (accessed on 5 September 2019).
3. Multi-Agency Radiation Survey and Site Investigation Manual (MARSSIM). (NUREG-1575, Revision 1). Available online: https://www.nrc.gov/reading-rm/doc-collections/nuregs/staff/sr1575/r1/ (accessed on 5 September 2019).
4. Sullivan, P.O.; Nokhamzon, J.G.; Cantrel, E. Decontamination and dismantling of radioactive concrete structures. *NEA News* **2010**, *28*, 27–29.
5. Dounreay Particles Advisory Group. Available online: https://assets.publishing.service.gov.uk/government/uploads/system/uploads/attachment_data/file/696380/DPAG_3rd__Report_September_2006.pdf (accessed on 5 September 2019).
6. Dennis, F.; Morgan, G.; Henderson, F. Dounreay hot particles: The story so far. *J. Radiol. Prot.* **2007**, *27*, A3–A11. [CrossRef] [PubMed]

7. Popp, A.; Ardouin, C.; Alexander, M.; Blackley, R.; Murray, A. Improvement of a high risk category source buried in the grounds of a hospital in cambodia. In *Proceedings of the 3th International Congress of the International Radiation Protection Association, Glasgow, UK, 14–18 May 2012*; pp. 1–10.
8. Maeda, K.; Sasaki, S.; Kumai, M.; Sato, I.; Suto, M.; Ohsaka, M.; Goto, T.; Sakai, H.; Chigira, T.; Murata, H. Distribution of radioactive nuclides of boring core samples extracted from concrete structures of reactor buildings in the fukushima daiichi nuclear power plant. *J. Nucl. Sci. Technol.* **2014**, *51*, 1006–1023. [CrossRef]
9. Shippen, B.A.; Joyce, M.J. Extension of the linear depth attenuation method for the radioactivity depth analysis tool(RADPAT). *IEEE Trans. Nucl. Sci.* **2011**, *58*, 1145–1150. [CrossRef]
10. Shippen, A.; Joyce, M.J. Profiling the depth of caesium-137 contamination in concrete via a relative linear attenuation model. *Appl. Radiat. Isot.* **2010**, *68*, 631–634. [CrossRef] [PubMed]
11. Adams, J.C.; Joyce, M.J.; Mellor, M. The advancement of a technique using principal component analysis for the non-intrusive depth profiling of radioactive contamination. *IEEE Trans. Nucl. Sci.* **2012**, *59*, 1448–1452. [CrossRef]
12. Adams, J.C.; Joyce, M.J.; Mellor, M. Depth profiling 137Cs and 60Co non-intrusively for a suite of industrial shielding materials and at depths beyond 50mm. *Appl. Radiat. Isot.* **2012**, *70*, 1150–1153. [CrossRef]
13. Adams, J.C.; Mellor, M.; Joyce, M.J. Determination of the depth of localized radioactive contamination by 137Cs and 60Co in sand with principal component analysis. *Environ. Sci. Technol.* **2011**, *45*, 8262–8267. [CrossRef] [PubMed]
14. Ukaegbu, I.K.; Gamage, K.A.A. A novel method for remote depth estimation of buried radioactive contamination. *Sensors* **2018**, *18*, 507. [CrossRef]
15. Ukaegbu, I.K.; Gamage, K.A.A. A model for remote depth estimation of buried radioactive wastes using CdZnTe detector. *Sensors* **2018**, *18*, 1612. [CrossRef] [PubMed]
16. Ukaegbu, I.K.; Gamage, K.A.A.; Aspinall, M.D. Nonintrusive depth estimation of buried radioactive wastes using ground penetrating radar and a gamma ray detector. *Remote Sens.* **2019**, *11*, 141. [CrossRef]
17. Knoll, G. Radiation interactions. In *Radiation Detection and Measurement*, 4th ed.; John Wiley and Sons Inc.: Hoboken, NJ, USA, 2010; Chapter 2; pp. 47–53.
18. Wagenmakers, E.-J.; Lee, M.; Lodewyckx, T.; Iverson, G.J. Bayesian versus frequentist inference. In *Bayesian Evaluation of Informative Hypotheses*; Springer: New York, NY, USA, 2008; pp. 181–207. [CrossRef]
19. Bishop, C. *Pattern Recognition and Machine Learning*; Springer: Secaucus, NJ, USA, 2006; pp. 12–19, 48–58.
20. Kucukelbir, A.; Tran, D.; Gelman, A.; Blei, D.M. Automatic differentiation variational inference. *J. Mach. Learn. Res.* **2017**, *18*, 1–45.
21. Salvatier, J.; Wiecki, T.V.; Fonnesbeck, C. Probabilistic programming in python using PyMC3. *PeerJ Comput. Sci.* **2016**, *2*, e55. [CrossRef]
22. Kim, J.; Lim, K.T.; Kim, J.; Kim, C.; Jeon, B. Quantitative analysis of NaI (Tl) gamma-ray spectrometry using an artificial neural network. *Nucl. Instrum. Methods Phys. Res. Sect. A Accel. Spectrometers Detect. Assoc. Equip.* **2019**, *944*, 162549. [CrossRef]
23. Goorley, J.T.; James, M.R.; Booth, T.E.; Brown, F.B.; Bull, J.S.; Cox, L.J.; Durkee, J.W.; Elson, J.S.; Fensin, M.L.; Forster, R.A.; et al. Initial MCNP6 Release Overview—MCNP6 version 1.0. Available online: https://permalink.lanl.gov/object/tr?what=info:lanl-repo/lareport/LA-UR-13-22934 (accessed on 5 September 2019).
24. Customs, U.S.; Protection, B.; Nuclear, D.; Office, D. *Compendium of Material Composition Data for Radiation Transport Modeling*; Technical report; Pacific Northwest National Laboratory: Washington, DC, USA, 2011.
25. Jeon, B.; Kim, J.; Moon, M.; Cho, G. Parametric optimization for energy calibration and gamma response function of plastic scintillation detectors using a genetic algorithm. *Nucl. Instrum. Methods Phys. Res. Sect. A Accel. Spectrometers Detect. Assoc. Equip.* **2019**, *930*, 8–14. [CrossRef]
26. Kim, J.; Lim, K.T.; Kim, J.; Kim, Y.; Kim, H. Quantification and uncertainty analysis of low-resolution gamma-ray spectrometry using Bayesian inference. *Nucl. Instrum. Methods Phys. Res. Sect. A Accel. Spectrometers Detect. Assoc. Equip.* **2019**, in press. [CrossRef]

Review

Passive Gamma-Ray and Neutron Imaging Systems for National Security and Nuclear Non-Proliferation in Controlled and Uncontrolled Detection Areas: Review of Past and Current Status

Hajir Al Hamrashdi *, Stephen D. Monk and David Cheneler

Department of Engineering, Lancaster University, Lancaster LA1 4YW, UK; s.monk@lancaster.ac.uk (S.D.M.); d.cheneler@lancaster.ac.uk (D.C.)
* Correspondence: h.alhamrashdi@lancaster.ac.uk

Received: 28 April 2019; Accepted: 8 June 2019; Published: 11 June 2019

Abstract: Global concern for the illicit transportation and trafficking of nuclear materials and other radioactive sources is on the rise, with efficient and rapid security and non-proliferation technologies in more demand than ever. Many factors contribute to this issue, including the increasing number of terrorist cells, gaps in security networks, politically unstable states across the globe and the black-market trading of radioactive sources to unknown parties. The use of passive gamma-ray and neutron detection and imaging technologies in security-sensitive areas and ports has had more impact than most other techniques in detecting and deterring illicit transportation and trafficking of illegal radioactive materials. This work reviews and critically evaluates these techniques as currently utilised within national security and non-proliferation applications and proposes likely avenues of development.

Keywords: passive radiation detection; gamma-ray; neutron; illicit trafficking; national security; non-proliferation

---

1. Introduction

Due to the hazardous ionising and activating nature of neutron and gamma radiation, there is a requirement to control and monitor the radiological materials, which produce them. Neutron and gamma-ray detection can directly lead to the identification of radiological sources in general, including nuclear materials. Due to the potential of these materials to be developed into nuclear weapons, these substances can pose direct threats to national security, and so are of great interest.

Illicit trafficking of nuclear materials and other radiological sources present a global threat that international organisations such as the IAEA (International Atomic Energy Agency) are forced to tackle frequently [1,2]. The IAEA Incident and trafficking database reported 3235 confirmed incidents of nuclear and other radioactive materials out of regulatory control between 1993 and 2017. Of these incidents, 278 were associated with trafficking or malicious use of materials such as highly-enriched uranium, plutonium and plutonium–beryllium neutron sources [2]. This issue highlights the importance of the effective control of nuclear and radiation materials at national and international cross points such as borders, ports and airports.

Effective application of radiation detection techniques requires knowledge of the environment in which the technology will be implemented, and the associated circumstances. In a controlled detection area such as an airport checkpoint, border line checkpoint, cargo inspection checkpoint or air cargo inspection, the space, and in most cases the physical contact time, allow for a reasonable level of flexibility. In an uncontrolled detection area such as buffer zones, airports terminals, train stations and public roads, space and physical contact time are less flexible and require more advanced detection technologies [3].

This review compares the various technologies utilised in radiation portal monitoring (RPM) of illicit radioactive materials including radiation sources, by-product materials and nuclear materials, with a view of identifying their advantages and limitations.

2. Radioactive Materials, Nuclear Materials and Radiation Sources:

Radioactive materials are defined by the IAEA as materials being designated in the national law or by a regulatory body as being subject to regulatory control because of their radioactivity [4]. Nuclear material is similarly defined as:

- Any plutonium isotope concentration except that with 80% or more of ^{238}Pu,
- Uranium enriched in the isotopes ^{233}U or ^{235}U,
- Uranium containing the mixture of isotopes as occurring in nature other than in the form of ore or ore-residue,
- Any material containing one or more of the above [4].

A radiation source is usually defined as artificially refined radioactive material produced outside the nuclear fuel cycles of research and power reactors [4,5]. The choice of radiation detection technology employed is primarily based on the radiation type being emitted, the amount of radiation, the energy spectra and whether the radioactive isotope needs to be identified. Predominantly, nuclear security-based applications are interested in detecting either gamma-rays (typically E > 10 keV), and/or neutrons [6–8]. Gamma-rays are typically emitted from an excited nucleus going from a higher energy state to a lower energy state, usually following the decay of its parent nucleus. Several mechanisms, such as fission and fusion reactions, neutron capture reactions, annihilation reactions and activation processes, can all result in the emission of gamma-rays. Because gamma-ray assay and spectra measurements are the easiest and most common technologies, they are of tensed to identify and differentiate different nuclear materials and their isotopic composition [7]. Figure 1 shows the gamma-ray intensity spectra and characteristic peaks for various nuclear material isotopes [7,9,10].

Other gamma emitting radiation sources that are often found to be involved in illicit trafficking are ^{192}Ir, ^{137}Cs and ^{241}Am [2]. Figure 2 shows gamma-ray characteristic peaks of these three isotopes.

Neutron emission detection and neutron assay is another common procedure used to detect and identify nuclear materials and radiation sources [6,7]. Neutron sources in nature and industry can be categorised as spontaneous fission sources, reactor sources, alpha-neutron sources, photo-neutron (or gamma-neutron) sources and ion accelerator sources as shown in Table 1 [6,11–13].

Table 1. Neutron sources and average energies.

Neutron Source	Neutron Source Type	Average Neutron Energy (MeV)	Half-Life (Years)
^{252}Cf	Spontaneous fission	1–3 (2.35 [1])	2.645
^{241}Am-9Be	Alpha-neutron source	4.2	432.2
^{239}Pu-9Be	Alpha-neutron source	4–5	24,114 years
^{124}Sb-9Be	Photo-neutron source	0.025 (close to mono-energetic)	0.164 (60 days)
D-D reaction	Accelerator source	2.4 (close to mono-energetic)	N/A
D-T reaction	Accelerator source	14.1(close to mono-energetic)	12.32

[1]: Reference [8], page 93.

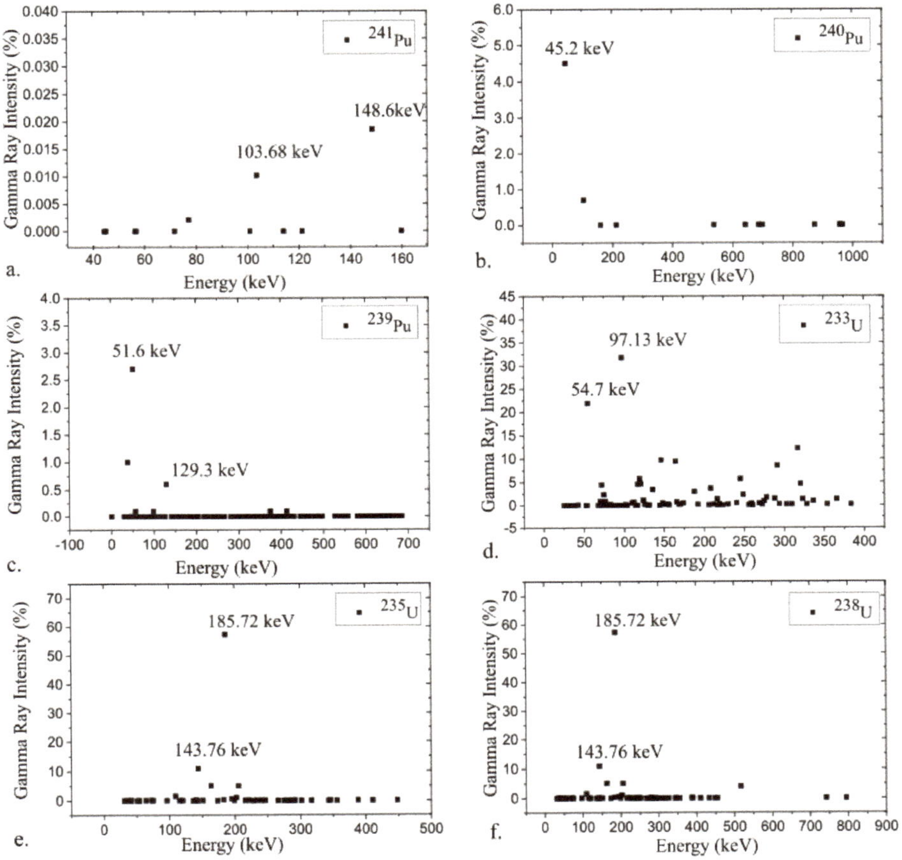

Figure 1. Characteristic gamma spectrum and gamma peaks of nuclear materials isotopes a. [241]Pu, b. [240]Pu, c. [239]Pu, d. [233]U, e. [235]U, f.[238]U (Data source: Idaho National Engineering and Environmental Laboratory [9]).

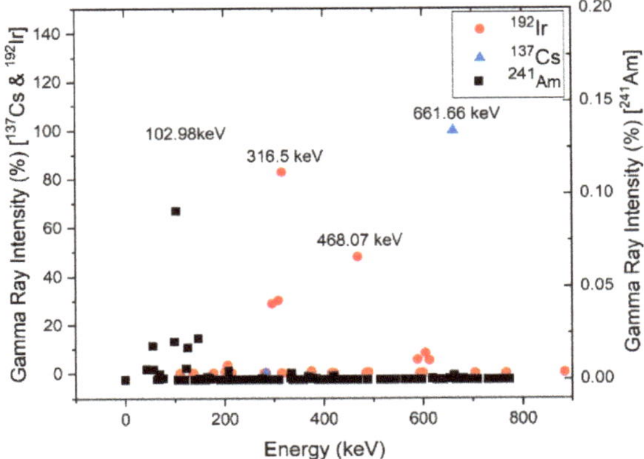

Figure 2. Gamma-ray characteristic energies and energy peaks of ^{192}Ir, ^{137}Cs and [241]Am.

Production of tritium from accelerator-based sources is affected by the closure of tritium-production reactors, non-proliferation policies and funding cuts. Other sources of tritium are breeding redactions in lithium blankets [14]. Other possible sources of neutron are D_Li-reactions [15] and spallation reactions [16]. Neutron multiplicity $\widetilde{v}$, or the number of neutrons emitted per fission, is a parameter obtained in the result of an analysis or measure. Table 2 gives a list of spontaneous fission isotopes commonly subjected to neutron multiplicity assays [6,7,10,17].

Table 2. Spontaneous fission isotopes and neutron multiplicity.

Isotope	Neutron Number	Total Half-Life (Years)	Average Spontaneous Fission Multiplicity
^{242}Cm	146	0.447	2.528
^{249}Bk	152	0.877	3.4
^{252}Cf	154	2.645	3.768
^{248}Cm	148	3.84	3.161
^{240}Pu	146	6.56	2.151
^{238}Pu	144	87.7	2.21
^{238}U	143	4.47×10^9	2.0
^{235}U	146	7.04×10^8	1.87

Induced fission multiplicity depends on the fission isotopes and the energy of the incident neutrons [17,18]. Figure 3 illustrates neutron spectrum multiplicity for nuclear materials ^{235}U and ^{239}Pu as functions of energy.

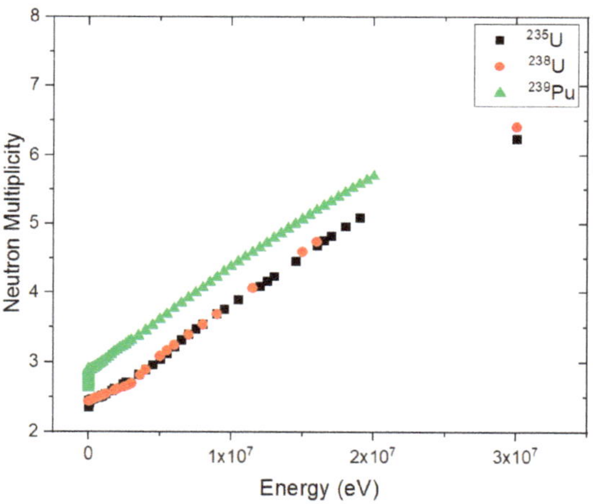

Figure 3. Neutron multiplicity as function of incident neutron energy for ^{235}U and ^{239}Pu.

Unlike gamma-rays, the wide energy spectrum of emitted neutrons, and the change in their energy as they traverse materials, make source identification through the energy of emitted neutrons a less effective method of detection. However, the increasing volume of research in this field such as the research in the field of neutron scattering cameras may indicate the emergence of new technologies [19–23].

3. Problem Definition and Authorities' Requirements

The major concern involving illicit trafficking and proliferation of nuclear materials is the threat of using these materials in criminal activities and terrorist acts. This concern has been gradually increasing during the last three decades and is becoming a definite threat in times of international instability and travel. The subject of illegal nuclear trafficking and unlawful nuclear acts is becoming the primary concern of international and global agencies such as the IAEA [24], European Commission [25] and Interpol [26]. Other factors including the economic and political impacts of this illicit trafficking are also part of the multithreaded problem.

As with many illegal acts at the international level, security plans and prevention policies along with international legislation have been implemented to deter and prevent illicit trafficking and to promote nuclear non-proliferation. Examples of these plans and treaties are the Treaty on Non-proliferation of Nuclear Weapons in 1970 [27] and IAEA safeguards agreement and Code of Conduct on the Safety and Security of Radioactive Sources in 2004 [28]. Another example of international cooperation to deter illicit trafficking of nuclear materials is demonstrated by the adoption of the practices espoused in the Handbook of Nuclear Law produced by the IAEA [29]. This is the result of international organisations assisting legislation and regulatory bodies in member states in creating a strong and robust regulatory framework [29]. Other international, regional and cross continent agreements such as the International Convention for the Suppression of Acts of Nuclear Terrorism (ICSANT) [30] are part of the global effort to combat and prevent illicit trafficking of nuclear and radiological materials.

The safeguarding of radioactive materials in general is a continuous process, from the generation stage to the decommissioning stage, especially for nuclear materials. The uninterrupted tracking of these materials is the optimal method to safeguard and diminish the possibilities of illegal trafficking. While the situation norm is the controlled and legal transport of radioactive materials, incidents are still reported [2]. A series of protocols and procedures have been implemented at the national and international level to prevent these incidents. One of the most important factors in this process is the implementation of the means of detecting, identifying and localising radioactive materials using radiation detection equipment and radiation imaging techniques.

The main purpose of implementing detection and imaging technologies in these applications is the timely and accurate identification of illegal acts and the generation of evidence to enforce legal proceedings to eliminate trafficking networks [24,31]. The implementation of radiation detection and imaging technologies varies from state to state, but these technologies are generally implemented on sites where radioactive sources' life cycles are spent, such as nuclear reactors, hospitals, etc., and at national and international cross borders. Many parameters affect the efficacy of radioactive material detection, with the main factor being the performance of the technologies employed, especially their ability to identify and localise radioactive sources [32]. Other directly related parameters that can influence the choice of technology employed are the field of view, the potential targets and the time constraints. The area of interest is the location where the detection or imaging instrument will be stationed and the zone that needs to be monitored. As implied in the Introduction, this area can be categorised as controlled or uncontrolled and varies in terms of the size of the area to be scanned, the detector to source distance, the number of people/vehicles/items to be monitored and the extent of the shielding or obstructions in the vicinity. The nature of the potential targets affects the choice of detection or imaging system due to their inherent shielding characteristics, i.e., nuclear material hidden inside the engine block of a large truck will be difficult to detect from a distance due to the significant shielding this environment affords. In addition, regulations that preclude the use of active interrogation systems on targets for health and safety reasons may also affect the selection process, if scanning pedestrians or queues of passengers, for instance. Timing is another parameter that affects the selection process. Controlled areas such as airports and land ports are busy areas. For example, the daily average number of people at a busy airport like Heathrow Airport is over 200,000 passengers per day [33]. There will be a limit to how long passengers can be held for security checks for logistical reasons. Therefore, detection efficiency, data analysis speed and spatial resolution are key aspects of the specification of the technologies employed. The size of the detection or imaging system can as well be seen as a factor on the selection process. Pocket-type instruments are used to detect the presence of radioactive materials and in some cases the radiation level, usually to calculate personal dose. Hand-held instruments have higher sensitivity and can be used to detect, locate and characterise radioactive sources. Finally, fixed and vehicle-based devices are usually used at borders cross-points, seaports and similar controlled areas [32].

The IAEA suggests that there are over a hundred different forms of non-destructive analysis techniques available to be used in the process of identifying radioactive materials [31]. However, the most common detection and imaging devices utilise gamma-rays and/or neutrons. The specification of suitable gamma-ray and neutron detection equipment varies according to legislation and the safeguarding abilities of states. A set of criteria have been recommended by the IAEA in a collaboration with World Custom Organization (WCO), EUROPOL and INTERPOL. The main components in this set of recommendations are [31,32]:

Gamma-ray systems' requirements:

- At a mean dose rate of 0.2 μSv/h, the alarm of the system should be activated when the dose rate increases in a period of 1 s by 0.1 μSv/h for a pocket size instrument, by 0.05 μSv/h for a handheld instrument and 0.1 μSv/h for a fixed-installation instrument, for a duration of one second with 99% detection accuracy.
- False alarm rate should be minimal, with background measures of 0.2 μSv/h, with a false alarm rate of less than one every 12 h for pocket size instruments, less than six per hour for handheld instruments and less than one per day for fixed-installation instruments.

Neutron systems requirements:

- The alarm of the system should be activated above a threshold of 20,000 n/s with a source to detector distance of 0.25 m for handheld instruments and 20,000 n/s in 5 s with source to detector distance of 2.0 m for fixed-installation instruments, using a system with 99% detection accuracy.
- False alarm rate should be minimal with less than six per hour for handheld instruments and one per day for fixed-installation instruments.

Similarly, the American National Standard for Evaluation and Performance of Radiation Detection Portal Monitors for Use in Homeland Security have a set of criteria for gamma-ray and neutron equipment; however, the set of requirements are relative to initial reference settings within the equipment [34]. Applying these requirements might limit direct implementation and might affect the response of the system. Test and Evaluation Capabilities and Methodologies Integrated Process Team (TECMIPT) Test Operations Procedures (TTOP) For Radiation Detection Systems—Specific Methods specifies the minimum performance requirements for gamma-ray and neutron detection instruments [35]. These specifications have direct implementation and offer detailed requirements relative to the size category of the system.

Gamma-ray systems' requirements:

- The alarm of the system should be activated when the count increases above the background level by 0.5 μSv/h in 2 s for Radionuclide Identification Devices (RIDs) in the pocket and handheld size categories.
- The alarm of the system should be activated with ^{232}Th, ^{137}Cs, and ^{133}Ba, ^{60}Co and ^{57}Co sources moving past the system at a speed of 2.22 m/s and distance of closest approach of 3 m for RIDs in the fixed installation size category.
- False alarm rate should be minimal with less than one every 10 h for pocket size and handheld instruments and less than one every two hours for fixed-installation size instruments.

Neutron systems requirements:

- The alarm of the system should be activated when the exposure is above the threshold of 20,000 n/s in 2 s with ^{252}Cf sources with a source to detector distance of 0.25 m for RIDs in the pocket size and handheld size categories.

- For a moving ^{252}Cf source with activity of 20,000 n/s and moving past the system at a speed of 2.22 m/s at a distance of closest approach of 3 m, the system has to be able to detect the source with up to 1 cm steel or 0.5 cm of lead of shielding for RIDs in the fixed installation size category.
- False alarm rate should be minimal with less than one every 10 h for pocket size and handheld instruments and less than one every two hours for fixed-installation size instruments.

4. Physical and Electronic Collimations

Neutrons and gamma-rays are uncharged high-energy radiation fields. Conventional converging and diverging techniques, as well as other optical techniques, are not applicable in this case. A device is needed to precisely identify the lines along which detected radiation fields are generated. Collimation is the key word here. Collimation of incident radiation can be done physically and/or electronically. Physical collimation and electronic collimation are well-established imaging techniques in the field of radiation detection. The basic concepts of each of these two collimation technique are discussed in this section.

4.1. Physical Collimation

Physical collimators are patterns of highly attenuating materials positioned in front of a detector to limit the direction of incident radiation quanta to specific directions. As a result, a shadow image is formed on the detector resulting in greatly improved spatial resolution. However, this approach causes a noticeable decline in the efficiency of the system since it limits the number of detectable radiation quanta [36]. Physical collimation for gamma-rays is more effective at lower energies as the probability of penetration through matter increases with gamma-ray energies above the energy peak of Compton scattering.

The simplest physical collimator design is the pinhole collimator, which consists of a single small aperture. This technique offers excellent angular resolution; however, it limits the geometrical efficiency of the system. Parallel holes collimator, converging and diverging collimators are arrays of opaque and transparent photon channels used in imaging where the system scans across the entire field of view. The technique improves the angular resolution of the system and slightly increases the solid angle. Figure 4 shows schematics of physical collimators types.

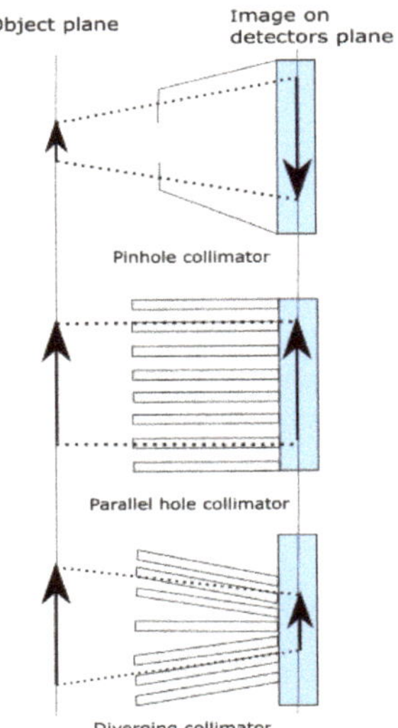

Figure 4. Schematic of physical collimator types.

A coded aperture is an alternative and popular form of physical collimation that was originally proposed for astrophysics measurements. It was first analytically proven effective for imaging systems in 1968 [37,38]. Commonly based on a 50% open mask with a large number of randomly distributed pinholes lying in a parallel plane with the detector, the technique offers higher efficiency compared to previously mentioned collimation techniques. Figure 5 illustrates the basic parameters of coded aperture imaging systems.

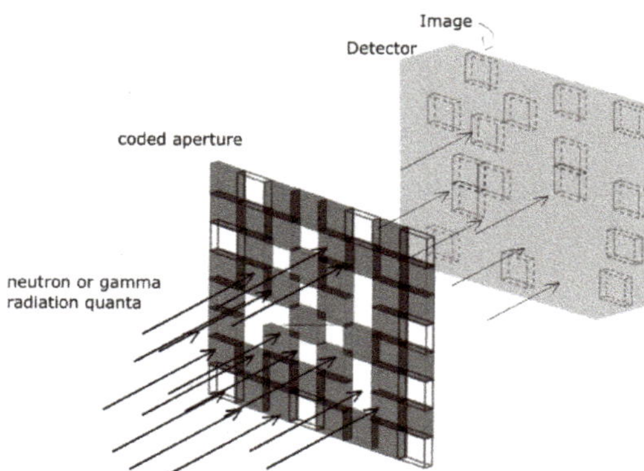

Figure 5. A schematic of coded aperture imaging system, with a coded aperture mask that is made with a pattern of opaque and open cells of highly attenuating materials followed by a radiation sensitive detector. Incident radiation field is attenuated in the coded aperture mask, with only a fraction of incident radiation is transmitted and detected on the system (based on reference [37]).

Coded aperture masks have greatly evolved since their inception, and, in most cases, their design can be tailored to fit the application requirements. There are generally two types of coded apertures: passive masks and active masks. In the case of passive masks, a highly absorbing material is used to stop and eliminate non-normally directed radiation quanta from reaching the detector. The choice of materials in passive masks mainly depends on the type of target radiation. High density/atomic number materials such as lead, tungsten and depleted uranium are often used to block high energy photons, while neutron absorbing materials such as high-density polyethylene (HDPE) and Gadolinium are used in coded apertures for neutron detection [39,40]. Passive physical collimation shows noticeable drawbacks over a considerable range of the energy spectrum, especially at high energies where radiation fields have enough energy to penetrate the opaque pattern of the mask [41–43]. On the other hand, active coded aperture designs use radiation sensitive materials, such as B- and Gd-doped glass plates for detecting low energy neutrons, as part of the collimation and detection process, which allows the detection of radiation quanta with a wider energy range [44–47]. Most of these active collimation examples combine physical collimation and Compton scattering in one system by using a pattern of large area detectors. Physical collimation is mainly utilised for detection of low gamma-ray energies, while Compton scattering is utilised for higher energy gamma-rays. Generally, the trade-off between angular resolution and detection efficiency is unavoidable in physical collimation. Higher activity sources or longer acquiring times (or both) are usually recommended to improve the efficiency of these systems.

4.2. Electronic Collimation: Compton Camera and Neutron Scattering Camera

Electronic collimation (widely known as a Compton camera for gamma-ray detection and neutron scattering camera for fast neutron detection) is a well-studied collimation approach, especially utilised within gamma-ray detection. Gamma Compton cameras are comprised of two pixelated detectors and utilise the laws of conservation of momentum and energy to infer the most probable trajectories of the scattered and/or absorbed radiation fields. The first detector scatters the gamma photon, which results in an electron being emitted and its energy measured. The second detector absorbs the scattered gamma photon and measures its energy. The location of the pixels activated in each detector determines the angle of scattering and hence the probable origin; the energy of the initial gamma photon can be calculated from measured energies of the incident and scattered photons [42,48]. The neutron scattering camera similarly utilises at least two detectors and the conservation of energy and momentum. However, in this instance, the reaction is between an incident fast neutron and a proton present in the proton-rich detectors in order to sense and localise the fast neutron source. The time-of-flight data of the scattered neutron is used to measure the energy of the incident neutron [19,23]. Figure 6 shows the basic elements in a two pixelated planes imaging system based on electronic collimation.

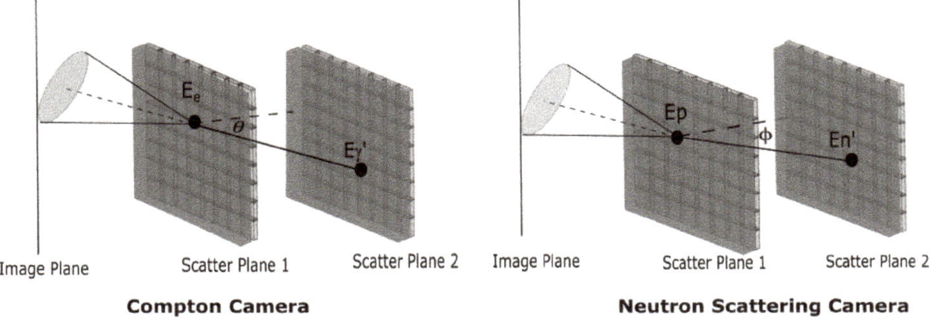

Figure 6. Schematic of basic parameters in Compton scattering camera (**left**) and neutron scattering camera (**right**).

5. Passive Detection Systems of Illicit Radioactive Materials

The two modes of detecting nuclear materials and other radioactive sources are mainly active mode and passive mode. Active detection mode (not part of the work presented here) uses externally generated neutrons, gamma-rays or X-rays to interrogate radioactive materials. This approach offers in-depth characterisation of target radioactive material, especially for fissile materials, although the major drawbacks are that it cannot be used in many circumstances, such as in proximity with humans and in uncontrolled detection areas [6,49–52].

In passive detection, an imaging system is used to detect and characterise neutrons and gamma-rays directly emitted from nuclear materials and radioactive sources. In contrast to an active detection technique, passive detection requires less architecture arrangement and conceivably lower in cost. In Safeguards Techniques and Equipment series by IAEA, approximately all gamma-ray non-destructive equipment discussed in the report are in passive mode [31]. In the same report, the ratio of listed passive to active neutron assay equipment is 4:1. This clearly shows the impact of passive detection mode at the international level in safeguard and security applications. A common design is the Radiation Portal Monitor (RPM), which typically consists of several detectors designed in a rectangle shape located at a fixed site [5]. Some passive imaging systems can characterise the radioactive material, reject background radiation and estimate the source to system distance. Passive detection systems offer a safe and simple detection mode, although the drawback is that its absolute efficiency decreases with increasing shielding around the radioactive material [53]. Since passive detection depends exclusively on the radioactive source under investigation and the detection system used, the statistical quality of results and the time to detect a source of specified strength depends mainly on characteristics such as intrinsic efficiency, angular resolution, spatial resolution and time resolution, shielding, and source-detector distance.

In this work, passive detection and identification systems are categorised based on the target radiation field: gamma-rays, neutrons and dual systems. In each category, the systems will be further classified into pocket-type instruments, hand-held instruments and large fixed or vehicle based instruments [54]. Another equally important classification factor is the purpose of detection instruments summarised as detection, assessment and localisation, and identification [54]. The following review attempts to compare and appraise past and present passive detection systems and techniques found in the literature that have been predominantly designed to detect and deter illicit trafficking, smuggling and transporting of nuclear materials and other radioactive sources.

5.1. Gamma-Ray Detection Systems

Common single crystal inorganic gamma detectors such as NaI, CsI, SrI2(Eu) and PVT (polyvinyltoluence organic scintillation detectors) or CdZnTe and HPGe (High Purity Germanium semiconductor detectors) are popular due to their stable performance, high efficiency and relatively low price [31,55–58]. NaI(Tl) is by far the most studied and most commercially successful inorganic scintillator [8]. However, single crystal imaging systems are far more sensitive to background radiation and are more prone to false alarms [3,59]. Pairing single crystal detector with signal analysers, such as multichannel analysers, might widen the scope of applications for this group of detectors [31]. However, imaging is almost always desirable, alongside detection, to enhance a system's sensitivity, angular resolution, energy resolution and localisation of point-like sources [60,61].

Physical collimation, in particular coded apertures, and Compton scattering techniques have both been adopted to enhance and improve the detection abilities of gamma imaging systems. Fixed installation coded aperture systems offer long distance and large area coverage with improved signal-to-background ratio [62,63]. However, these systems are best implemented at border controls, as they require fixed or slowly moving targets. Problems and limitations, such as false alarms and timing issues, as well as proposed solutions for this technology, such as energy windows and baseline suppression, are frequently discussed in literature [64–70]. Hand-held coded aperture systems offer a flexible solution for detecting and localising of radioactive materials [71,72]. In addition to the main goal of detecting and localising radioactive sources while scanning vehicles, people, luggage and cargo, other applications such as monitoring the extent of nuclear related emergencies have been suggested. Many mechanically collimated systems have found success in this field [72–75]. Table 3 summarises coded aperture-based gamma imaging systems found in the literature, including their detection method, their size category and the purpose of application.

Table 3. Coded aperture-based gamma-imaging systems.

System Size Definition	Examples and Proposed Application in Literature	Detector/s	Industrial Designation
Fixed installation	Detection and localisation [62,63]	CsI(Na)	
Fixed installation	Detection, assessment and localisation [76]	HPGe & NaI	MISTI
Fixed installation	Detection, assessment and localisation [75]	CdZnTe	ORIGAMIX
Fixed installation/hand-held	Detection and localisation [77]	NaI	RMC
Fixed installation/hand-held	Detection and localisation [73]	CsI(Tl)	CARTOGAM
Fixed installation	Detection, assessment and localisation [78]	(GSO)	
Hand-held	Detection and localisation [71]	CdTe-Medpixi2	
hand-held	Detection and localisation [74]	CsI(Na)	RADCAM
hand-held	Detection and localisation [79]	CsI(Tl)	
hand-held	Detection and localisation [72]	CdZnTe-Timepix	GAMPIX

In the energy range of nuclear material gamma-ray sources (60 keV to 3.0 MeV), Compton scattering is the dominant photon interaction mechanism, which makes the Compton scattering technique the most appropriate technique compared to other techniques [3]. Compton based systems feature a wide field of view with improved detection efficiency compared to mechanically collimated gamma imaging systems, especially for high-energy gamma-rays [3,80]. In addition, Compton systems offer the ability to detect, assess and localise a gamma-ray source with an associated reduction in background radiation [81]. Fixed installation and portable Compton systems are the most common size categories [82–85]. The performance of these systems varies between detectors with some using low energy resolution, high sensitivity NaI(Tl) and CsI(Tl) scintillations [82,83], while others use high resolution Si and HPGe semiconductor detectors [84]. Image reconstruction methods for Compton systems, such as Maximum Likelihood Expectation Maximization (MLEM), Maximum Likelihood Ratio (MLR) and stochastic origin ensembles, have been regularly studied and optimised for their direct impact on the performance of Compton system in this field [86,87]. There are hybrid-imaging systems that utilise both Compton camera and coded aperture technology; examples include passive mask [88] and active mask [45,89,90] systems. The duality in imaging techniques aims to utilise the advantages of both physical collimation and Compton scattering. However, designs need to take into account the optimum arrangement of layers to avoid negating these advantages. Another promising technique in gamma-ray imaging are 3D systems that utilise coded apertures or Compton scattering. The 3D systems are used in assessing and localising smuggled and hidden sources by projecting a 3D image of the search scene, which allow faster and easier navigation in the area of interest [91,92].

5.2. Neutron Detection Systems

Although most nuclear materials emit either or both neutron and gamma-rays, heavy shielding of gamma-rays can greatly lower the efficiency of gamma-ray imaging systems, negatively impacting their efficacy in nuclear materials' non-proliferation and safeguard applications. Neutrons are highly penetrating and nuclear materials emitting neutrons require bulky shielding to completely conceal neutrons. Therefore, neutron imaging systems are extensively used in nuclear materials imaging and they offer an excellent alternative. Due to their high thermal neutron cross section (5330 barns) and low gamma-ray sensitivity, ^{3}He gas filled counters have been the standard neutron monitoring technology for decades [31,93]. Thermal neutrons detection efficiency for ^{3}He gas filled counters is a function of the amount of ^{3}He gas and increases with increasing pressure. For example, a 72 in in height and 2 in in diameter ^{3}He tube under 3 atm pressure has efficiency of 3.05 cps/ng ^{252}Cf. The main supply of ^{3}He is the ^{3}H purification process, which has seen a dramatic decrease in the last two decades [93,94]. This has led to a continuous search for alternative neutron detection technologies. Direct gas filled counter alternatives such as BF_3 proportional counters, boron lined proportional counters and fission chambers have been commercially in use [31,95–97], but they have been shown to be significantly less efficient [94,97,98].

Neutron sensitive scintillation detectors and semiconductor detectors are frequently used in neutron detection. Neutron sensitive scintillation detectors include liquid and plastic organic scintillators [99–104], glass scintillators [105–108] scintillating fibres [109] and bubble chambers [110]. Bonner spheres are examples of radiation detectors embedded in a spherical moderator layer. Bonner spheres are well-established neutron spectrometer instruments in the field of nuclear dosimetry and inspection non-proliferation [111,112]. However, Bonner spheres have inherently low energy resolution and inverse relationship between moderator thickness and detection efficiency. Semiconductor based detectors are a less popular means of neutron detection due to their lower efficiency compared to the scintillation detection materials in this field and the occasional requirement of having to use foils or coatings of conversion material to convert neutrons into a detectable signal, usually electrons [13,113]. However, their ruggedness and high-speed response make them an interesting option for safeguarding and security applications [114,115]. Semiconductor materials such as ^{4}H-SiC, diamond and CdZnTe have been investigated in literature for their applications in neutron detection [116–119]. ^{4}H-SiC and SiC semiconductors are promoted for their abilities to work in high temperature and high radiation environments along with other desirable properties such as high energy band gap and lower production cost, compared to diamond, which has similar properties [116]. Diamond materials, such as diamond high pressure, high temperature (HPHT) synthetic diamond or diamond grown using CVD, are mechanically durable and inherently radiation hard. Like SiC detectors, diamond detectors have a wide band gap, which makes them highly appealing for radiation detection applications at high temperatures [118,120]. CdZnTe with neutron converting layer such as Gd are proposed for portable thermal neutron detection systems [119,121]. Activation foils were suggested for safeguard applications such as practical neutron flux measurement tools [122].

As for gamma-ray detection, collimation techniques in neutron detection are deployed to enhance detection efficiency, angular and energy resolutions, increase the field of view and decrease the acquisition time. In addition, for screening vehicles, cargo and large containers, the imaging systems should be accurate with a low probability of false alarms and low sensitivity to gamma-rays [123]. A number of simulation-based studies discuss potential neutron imaging systems with physical collimation or neutron scattering/ToF (Time of Flight) based collimation [124–127]. An equally important aspect in nuclear materials detection is the discrimination method used to discriminate between neutrons and gamma-rays [128]. Because gamma-rays are almost always present in the background, discrimination methods are crucially important and have been extensively studied in literature [129–134]. A range of radiation detection and identification systems are commercially available from vehicle size [135] to handheld size [136–138]. For a more detailed review of portal radiation monitors, Table 4 lists all neutron-imaging systems discussed and experimentally evaluated in literature for safeguard and non-proliferation of nuclear materials. The categorisation of proposed applications and the size definitions are based on those mentioned in Section 4.

5.3. Dual Gamma-Ray and Neutron Detection Systems

Dual particle imaging systems detect gamma-rays and neutrons simultaneously and can differentiate between the two radiations. This method of imaging has an advantage over single particle imaging methods because it allows the passive detection and identification of a wide range of nuclear materials and other radioactive sources.

There are two main groups of systems in the field of dual particle imaging. The first group is comprised of single materials that are sensitive to both gamma-rays and neutrons. The second group uses multiple detection materials systems with detectors not necessarily sensitive to both particles. The latter imaging technique offers a reduction in system complexity as additional discrimination techniques are not necessarily required. In addition, this category offers higher design flexibility, as the parameters employed to enhance system response to one radiation field are usually independent of the other.

Table 4. Examples of neutron imaging systems used in nuclear materials security, their collimation technique, properties and characteristics.

Author, Year and Reference	Proposed Application	Collimation/Detection Technique	System Size Definition	Main Detection Materials	Approximate Intrinsic Efficiency (Thermal Neutrons/Fast Neutron 252Cf) (%)
Miller et al. (2003) [139]	Detection, assessment and localisation	Neutron scatter	Fixed installation	Plastic scintillator	NA/NA
Bravar et al. (2006) [140]	Detection and assessment	Neutron scatter	Fixed installation	BC-404 plastic scintillator	NA/NA
Vanier et al. (2007) [141]	Detection, assessment and localisation	Neutron scatter	Fixed installation	Plastic scintillator	NA/NA
Mascarenhas et al. (2009) [142]	Detection, assessment and localisation	Neutron scatter	Fixed installation	EJ-301	NA/NA
Siegmund et al. (2009) [143]	Detection	Coded aperture and Stack of microchannel plates	Fixed installation	^{10}B doped microchannel plates	~20%/NA
Herbach et al. (2010) [144]	Detection and assessment	Null/gamma from neutron capture	Fixed installation	BGO with Cd converter	45%/NA
Ryzhikov et al. (2010) [145]	Detection and assessment	Null/gamma from neutron capture	Fixed installation	$CdWO_2$	67%/42%
Marleau et al. (2010) [146]	Detection and assessment	Active coded aperture	Fixed installation	EJ-301	NA/NA
Nakae et al. (2011) [147]	Detection and assessment	Null/Array of liquid scintillator	Fixed installation	Organic liquid scintillator (not specified)	NA/~6% (absolute)
Bellinger et al. (2012) [148]	Detection and assessment	Null/array of slabs	Hand-held	Si diodes with ^{6}LiF	6.8%/NA
Ide et al. (2012) [149]	Detection, assessment and localisation	Neutron scatter	Fixed scintillator	EJ-309	NA/NA
Joyce et al. (2014) [150]	Detection and assessment	Null/Multiplicity assay	Fixed scintillator	EJ-309	NA/
Brennan et al. (2015) [151]	Detection, assessment and localisation	Coded aperture and Time-encoded imaging	Fixed installation	Organic liquid scintillator (not specified)	NA/NA
Fronk et al. (2015) [152]	Detection and assessment	Null/double sided Microstructure	Hand-held	Si diodes with ^{6}LiF	~29.48%/NA
Ianakiev et al. (2015) [153]	Detection and assessment	Null/^{6}Li embedded in PVT	Fixed installation	^{6}Li and PVT	NA/NA
Hoshor et al. (2015) [154]	Detection and assessment	Null/array of slabs	Hand-held	Si diodes with ^{6}LiF	~22%/~4.5%
Goldsmith et al. (2016) [155]	Detection, assessment and localisation	Neutron scatter	Fixed installation	EJ-309	NA/45%
Fulvio et al. (2017) [156]	Detection and assessment	Ring of multiplicity counters	Fixed installation	EJ-309	NA/
Cowles et al. (2018) [157]	Detection and assessment	Null/multiple panels	Fixed installation	LiF/ZnS	36%/NA
Ochs et al. (2019) [158]	Detection	Microstructure semiconductor	Wearable device	Si diode with ^{6}LiF	30%/NA

Materials sensitive to both gamma-ray and neutron have been investigated for their dual detection abilities since the 1950s [159,160]. Examples of the list of detection materials range from inorganic scintillators [161–163], semiconductor detectors [164–166], glass organic scintillators [167], some classes of elpasolite scintillators [168,169], some classes of liquid scintillators and plastic scintillators [170–174]. The common feature between these detection materials is their superior ability to enable the distinguishing of gamma-ray signals from neutron signals by methods such as pulse shape discrimination and pulse height discrimination [133,134,170,175]. A handful of fixed installation and portable monitoring systems are suggested for security and non-proliferation applications are found in the literature. The scintillation materials used in these systems vary dramatically with ^{6}Li(Eu) and Li-glass detectors, EJ-309 liquid scintillators and CLYC (Cs_2LiYCl_6:Ce) elpasolite detectors being popular [162,176–178] with some also utilising coded aperture collimation to enhance imaging characteristics [179,180]. A number of examples of hand-held and pocket size systems for monitoring purposes similarly exist; almost all detection materials in this size category are based on plastic scintillators or elpasolite scintillation materials [99,181–184]. These systems offer flexibility and fast response, albeit with a limited field of view.

Since 2004, the research on multiple detection materials imaging systems for security and non-proliferation applications has increased. System abilities vary according to the detection and collimation method and the size of the system. Table 5 presents a timeline of multiple detectors imaging systems discussed in literature along with their collimation and detection techniques between 2004 and 2016.

Table 5. Timeline of dual particle multiple detectors imaging system in security and non-proliferation applications.

Year	Author and Reference	Collimation	Main Detection Materials
2004	Aryaeinejad and Spencer [185]	None	^{6}Li and ^{7}Li-loaded glass scintillators
2007	Baker et al. [186]	None	NaI(Tl) and LiI(Eu)
2008	Enqvist et al. [187]	None	Cross correlation BC-501A
2009	Runkle et al. [188]	None	NaI(Tl) and ^{3}He
2011	Polack et al. [189]	Compton and neutron scattering	NaI(Tl) and EJ-309
2012	Cester et al. [190]	None	LaBr(Ce), NaI(Tl), NE-213 and ^{3}He
2013	Ayaz-Maierhafer et al. [191]	Coded aperture	CsI and EJ-309
2014	Poitrasson-Rivière et al. [192]	Compton and neutron scattering	NaI(Tl) and EJ-309
2016	Cester et al. [193]	Null	EJ-420, EJ-560 and EJ-299-33A

A number of research papers have been undertaken and investments have been made into large area coverage using a network of detectors. This concept has been around for over a decade [194–196]; however, the realisation of the advantages of this technique along with the advances in network and communication fields will lead to new developments in this area. Examples of network systems and algorithms in this field are the RAdTrac network system for gamma detectors, the particle filter algorithm for a network of gamma counters and the ROSD-RSD (Ratio of Squared Distance-Radiation Source Distance) algorithm method [197–200]. Other systems like identiFINDER S900 [201] and SmartShieldTM v2.0 [202] are commercially available for radionuclides identification and tracking.

6. Conclusions

Illicit trafficking of nuclear materials and radioactive materials is a cross-border problem that must be tackled globally. Robust and efficient detection equipment and radiation detection systems stand on the front line of defence against the acts of illicit trafficking. However, understanding the different parameters that affect the choice of detection equipment and/or radiation detection systems can greatly help with installing the most effective detection techniques. The parameters that have the most effect are (more details in Section 3):

- Security agencies and legislation bodies requirements,
- Areas under surveillance and place of implementation,

- Image quality requirement,
- Timing and speed requirements.

Once the main requirements are established, the options can then be investigated within detection and/or imaging techniques of gamma-ray sensitive systems, neutron sensitive systems or dual gamma-ray and neutron sensitive systems. Each technique has its advantages over the others and the main stage in planning to install a detection system that will positively contribute in deterring illicit trafficking is to investigate and study each implementation site individually.

Funding: This work is in part supported by the UK Engineering and Physical Sciences Research Council (EPSRC), via grant EP/S020411/1.

Conflicts of Interest: The authors declare no conflict of interest.

References

1. Zaitseva, L.; Hand, K. Nuclear smuggling chains. Suppliers, intermediaries, and end-users. *Am. Behav. Sci.* **2003**, *46*, 822–844. [CrossRef]
2. IAEA. *Inicdents and Trafficking Database (ITDB) Incidents of Nuclear and Other Radioactive Material Out of Regulatory Control*; International Atomic Energy Agency: Vienna, Austria, 2018.
3. Byrd, R.C.; Moss, J.M.; Priedhorsky, W.C.; Pura, C.A.; Richter, G.W.; Saeger, K.J.; Scarlett, W.R.; Scott, S.C.; Wagner, R.L. Nuclear detection to prevent or defeat clandestine nuclear attack. *IEEE Sens. J.* **2005**, *5*, 593–609. [CrossRef]
4. IAEA. *Safety Glossary Terminology Used in Nuclear, Radiation, Radioactive Waste and Transport Safety Version 2.0; Department of Nuclear Safety and Security*; International Atomic Energy Agency: Vienna, Austria, 2006.
5. *Glossary of Terms for Nuclear, Biological, and Chemical Agents and Defense Equipment*; U.S. Army Medical Department and U.S. Army Medical Department, Army Public Health Center: Washington, DC, USA, 2001.
6. Gozani, T. *Active Nondestructive Assay of Nuclear Materials, Principles and Applications*; US Nuclear Regulatory Commission: Washington, DC, USA, 1981.
7. Reilly, D.; Ensslin, N.; Smith, H.; Kreiner, S. *Passive Nondestructive Assay of Nuclear Materials*; National Technical Information Service, U.S. Department of Commerce: Washington, DC, USA, 1991.
8. Milbrath, B.D.; Peurrung, A.J.; Bliss, M.; Weber, W.J. Radiation detector materials: An overview. *J. Mater. Res.* **2008**, *23*, 2561–2581. [CrossRef]
9. *Gamma-Ray Spectrometry Catalog*; The Idaho National Laboratory: Idaho Falls, ID, USA, 1999.
10. Gozani, T. Fission signatures for nuclear material detection. *IEEE Trans. Nucl. Sci.* **2009**, *56*, 736–741. [CrossRef]
11. Stewart, L. Neutron Spectrum and Absolute Yield of a Plutonium-Beryllium Source. *Phys. Rev.* **1955**, *98*, 740–743. [CrossRef]
12. Krane, K.S. *Introductry Nuclear Physics*; John Wiley and Sons: New York, NY, USA, 1988.
13. Knoll, G.F. *Radiation Detection and Measurment*, 4th ed.; John Wiley and Sons: New York, NY, USA, 2010.
14. Yapıcı, H.; Şahin, N.; Bayrak, M. Investigation of neutronic potential of a moderated (D–T) fusion driven hybrid reactor fueled with thorium to breed fissile fuel for LWRs. *Energy Convers. Manag.* **2000**, *41*, 435–447. [CrossRef]
15. Knaster, J.; Arbeiter, F.; Cara, P.; Chel, S.; Facco, A.; Heidinger, R.; Ibarra, A.; Kasugai, A.; Kondo, H.; Micciche, G.; et al. IFMIF, the European–Japanese efforts under the Broader Approach agreement towards a Li(d,xn) neutron source: Current status and future options. *Nucl. Mater. Energy* **2016**, *9*, 46–54. [CrossRef]
16. Mansur, L.K.; Rowcliffe, A.F.; Nanstad, R.K.; Zinkle, S.J.; Corwin, W.R.; Stoller, R.E. Materials needs for fusion, Generation IV fission reactors and spallation neutron sources–similarities and differences. *J. Nucl. Mater.* **2004**, *329–333*, 166–172. [CrossRef]
17. Nea, O. Java-Based Nuclear Data Information System. Available online: http://www.oecd-nea.org/janis/ (accessed on 10 February 2016).
18. Terrell, J. Distributions of Fission Neutron Numbers. *Phys. Rev.* **1957**, *108*, 783. [CrossRef]
19. Herzo, D.; Koga, R.; Millard, W.A.; Moon, S.; Ryan, J.; Wilson, R.; Zych, A.D.; White, R.S. A Large Double Scatter Telescope for Gamma Rays and Neutrons. *Nucl. Instrum. Methods* **1975**, *123*, 583–597. [CrossRef]

20. Wunderer, C.B.; Holslin, D.; Macri, J.R.; McConnell, M.; Ryan, J.M. SONTRAC-a low background, large area solar neutron spectrometer. In Proceedings of the Conference on the High Energy Radiation Background in Space, Snowmass, CO, USA, 22–23 July 1997; pp. 73–76.

21. Ryan, J.M.; Desorgher, L.; Flückiger, E.O.; Macri, J.R.; McConnell, M.L.; Miller, R.S. SONTRAC: An imaging spectrometer for solar neutrons. In Proceedings of the Proceedings of SPIE—The International Society for Optical Engineering, Waikoloa, HI, USA, 11 February 2003; pp. 399–410.

22. Marleau, P.; Brennan, J.; Krenz, K.; Mascarenhas, N.; Mrowka, S. Advances in imaging fission neutrons with a neutron scatter camera. In Proceedings of the IEEE Nuclear Science Symposium Conference Record, Honolulu, HI, USA, 26 October–3 November 2007; pp. 170–172.

23. Mascarenhas, N.; Brennan, J.; Krenz, K.; Lund, J.; Marleau, P.; Rasmussen, J.; Ryan, J.; Macri, J. Development of a neutron scatter camera for fission neutrons. In Proceedings of the IEEE Nuclear Science Symposium Conference Record, San Diego, CA, USA, 29 October–1 November 2006; pp. 185–188.

24. IAEA. *Combating Illicit Trafficking in Nuclear and Other Radioactive Materials*; International Atomic Energy Agency, IAEA: Vienna, Austria, 2007.

25. JRC: 20 Years Combating Illicit Trafficking of Nuclear Materials. Available online: https://ec.europa.eu/jrc/en/news/jrc-20-years-combating-illicit-trafficking-nuclear-materials-7023 (accessed on 7 March 2019).

26. Radiological and Nuclear Prevention. Available online: https://www.interpol.int/en/Crimes/Terrorism/Radiological-and-Nuclear-terrorism/Radiological-and-nuclear-prevention (accessed on 7 March 2019).

27. Treaty on the Non-Proliferation of Nuclear Weapons (NPT). Available online: https://www.un.org/disarmament/wmd/nuclear/npt/text (accessed on 7 March 2019).

28. *Code of Conduct on The Safety and Secuirty of Radioactive Sources*; International Atomic Energy Agency, IAEA: Vienna, Austria, 2004.

29. *Handbook on Nuclear Law*; International Atomic Energy Agency, IAEA: Vienna, Austria, 2003.

30. International Convention for the Suppression of Acts of Nuclear Terrorism. *Nucl. Terror. Conv.* **2005**, *2445*, 89.

31. *Safeguards Techniques and Equipment*; International Atomic Energy Agency, IAEA: Vienna, Austria, 2011.

32. *Detection of Radioactive Material at Borders*; International Atomic Energy Agency, IAEA: Vienna, Austria, 2002.

33. Heathrow Our Company: Facts and Figures. Available online: https://www.heathrow.com/company/company-news-and-information/company-information/facts-and-figures (accessed on 12 March 2019).

34. *Detection of Radioactive Materials at Borders Radiation Safety Section*; Jointly sponsored by IAEA, WCO, EUROPOL and INTERPOL; International Atomic Energy Agency: Vienna, Austria, 2002.

35. *American National Standard for Evaluation and Performance of Radiation Detection Portal Monitors for Use in Homeland Security*; ANSI N42.35-2016 (Revision of ANSI N42.35-2006); ANSI: Washington, DC, USA, 2016; pp. 1–70. [CrossRef]

36. *TECMIPT Test Operations Procedures (TTOP) For Radiation Detection Systems–Specific Methods*; NIST: Gaithersburg, MD, USA, 2012.

37. Gormley, J.E.; Rogers, W.L.; Clinthorne, N.H.; Wehe, D.K.; Knoll, G.F. Experimental comparison of mechanical and electronic gamma-ray collimation. *Nucl. Instrum. Methods Phys. Res. Sect. A-Accel. Spectrometers Detect. Assoc. Equip.* **1997**, *397*, 440–447. [CrossRef]

38. Dicke, R.H. Scatter-Hole cameras for X-rays and gamma rays. *Astrophys. J.* **1968**, *153*, L101. [CrossRef]

39. Ables, J.G. Fourier transform photography: A new method for X-ray astronomy. *Publ. Astron. Soc. Aust.* **1968**, *1*, 172–173. [CrossRef]

40. Whitney, C.M.; Soundara-Pandian, L.; Johnson, E.B.; Vogel, S.; Vinci, B.; Squillante, M.; Glodo, J.; Christian, J.F. Gamma-neutron imaging system utilizing pulse shape discrimination with CLYC. *Nucl. Instrum. Methods Phys. Res. Sect. A-Accel. Spectrometers Detect. Assoc. Equip.* **2015**, *784*, 346–351. [CrossRef]

41. Zou, Y.B.; Schillinger, B.; Wang, S.; Zhang, X.S.; Guo, Z.Y.; Lu, Y.R. Coded source neutron imaging with a MURA mask. *Nucl. Instrum. Methods Phys. Res. Sect. A-Accel. Spectrometers Detect. Assoc. Equip.* **2011**, *651*, 192–196. [CrossRef]

42. Todd, R.W.; Nighting, J.; Everett, D.B. Proposed gamma camera. *Nature* **1974**, *251*, 132–134. [CrossRef]

43. Everett, D.B.; Fleming, J.S.; Todd, R.W.; Nightingale, J.M. Gamma-radiation imaging-system based on compton-effect. In *Proceedings of the Institution of Electrical Engineers-London*; IET Digital Library: London, UK, 1977; Volume 124, pp. 995–1000.

44. Singh, M.; Brechner, R.R. Experimental test-object study of electronically collimated SPECT. *J. Nucl. Med.* **1990**, *31*, 178–186.

45. Cunningham, M.F.; Blakeman, E.; Fabris, L.; Habte, F.; Ziock, K. Active-mask coded-aperture imaging. In Proceedings of the IEEE Nuclear Science Symposium/Medical Imaging Conference, Honolulu, HI, USA, 26 October–3 November 2007; pp. 1217–1221.

46. Schultz, L.J.; Wallace, M.S.; Galassi, M.C.; Hoover, A.S.; Mocko, M.; Palmer, D.M.; Tornga, S.R.; Kippen, R.M.; Hynes, M.V.; Toolin, M.J.; et al. Hybrid coded aperture and Compton imaging using an active mask. *Nucl. Instrum. Methods Phys. Res. Sect. A-Accel. Spectrometers Detect. Assoc. Equip.* **2009**, *608*, 267–274. [CrossRef]

47. Marleau, P.; Brennan, J.; Brubaker, E.; Hilton, N.; Steele, J. Active Coded Aperture Neutron Imaging. In Proceedings of the 2009 IEEE Nuclear Science Symposium Conference Record, Orlando, FL, USA, 24 October–1 November 2009; Volumes 1–5, pp. 1974–1977. [CrossRef]

48. Woolf, R.S.; Phlips, B.F.; Hutcheson, A.L.; Mitchell, L.J.; Wulf, E.A. An Active Interrogation Detection System (ACTINIDES) Based on a Dual Fast Neutron/Gamma-Ray Coded Aperture Imager. In Proceedings of the 2012 IEEE International Conference on Technologies for Homeland Security, Waltham, MA, USA, 13–15 November 2012; pp. 30–35.

49. Singh, M. An Electronically Collimated Gamma Camera for Single Photon Emission Computed Tomography. Part I: Theoretical Considerations and Design Criteria. *Med. Phys.* **1983**, *10*, 421–427. [CrossRef]

50. Runkle, R.C.; Chichester, D.L.; Thompson, S.J. Rattling nucleons: New developments in active interrogation of special nuclear material. *Nucl. Instrum. Methods Phys. Res. Sect. A-Accel. Spectrometers Detect. Assoc. Equip.* **2012**, *663*, 75–95. [CrossRef]

51. Norman, D.R.; Jones, J.L.; Haskell, K.J.; Vanier, P.E.; Forman, L. Active nuclear material detection and imaging. In Proceedings of the 2005 IEEE Nuclear Science Symposium Conference Record, Fajardo, Puerto Rico, 23–29 October 2005; Volumes 1–5, pp. 1004–1008.

52. Jones, J.L.; Norman, D.R.; Haskell, K.J.; Sterbentz, J.W.; Yoon, W.Y.; Watson, S.M.; Johnson, J.T.; Zabriskie, J.M.; Bennett, B.D.; Watson, R.W.; et al. Detection of shielded nuclear material in a cargo container. *Nucl. Instrum. Methods Phys. Res. Sect. A-Accel. Spectrometers Detect. Assoc. Equip.* **2006**, *562*, 1085–1088. [CrossRef]

53. Chichester, D.L.; Simpson, J.D.; Lemchak, M. Advanced compact accelerator neutron generator technology for active neutron interrogation field work. *J. Radioanal. Nucl. Chem.* **2007**, *271*, 629–637. [CrossRef]

54. Runkle, R.C.; Smith, L.E.; Peurrung, A.J. The photon haystack and emerging radiation detection technology. *J. Appl. Phys.* **2009**, *106*, 7. [CrossRef]

55. Guss, P.; Reed, M.; Yuan, D.; Beller, D.; Cutler, M.; Contreras, C.; Mukhopadhyay, S.; Wilde, S. Size effect on nuclear gamma-ray energy spectra acquired by different-sized CeBr3, LaBr3:Ce, and NaI:Tl gamma-ray detectors. *Nucl. Technol.* **2014**, *185*, 309–321. [CrossRef]

56. Reinhard, M.I.; Prokopovich, D.; Van Der Gaast, H.; Hill, D. Detection of illicit nuclear materials masked with other gamma-ray emitters. In Proceedings of the IEEE Nuclear Science Symposium Conference Record, San Diego, CA, USA, 29 October–1 November 2006; pp. 270–272.

57. Siciliano, E.R.; Ely, J.H.; Kouzes, R.T.; Milbrath, B.D.; Schweppe, J.E.; Stromswold, D.C. Comparison of PVT and NaI(Tl) scintillators for vehicle portal monitor applications. *Nucl. Instrum. Methods Phys. Res. Sect. A-Accel. Spectrometers Detect. Assoc. Equip.* **2005**, *550*, 647–674. [CrossRef]

58. Mortreau, P.; Berndt, R. Determination of 235U enrichment with a large volume CZT detector. *Nucl. Instrum. Methods Phys. Res. Sect. A-Accel. Spectrometers Detect. Assoc. Equip.* **2006**, *556*, 219–227. [CrossRef]

59. Sjoden, G.E.; Detwiler, R.; Lavigne, E.; Baciak, J.E., Jr. Positive SNM gamma detection achieved through synthetic enhancement of sodium iodide detector spectra. *IEEE Trans. Nucl. Sci.* **2009**, *56*, 1329–1339. [CrossRef]

60. Kim, K.H.; Jun, J.Y.; Jun, I.S.; Kwak, S.W. Development of a car-mounted nuclear material monitoring system: A prototype system. *Nucl. Instrum. Methods Phys. Res. Sect. A-Accel. Spectrometers Detect. Assoc. Equip.* **2009**, *607*, 154–157. [CrossRef]

61. TSA MD134. Available online: http://www.rapiscansystems.com/en/products/radiation_detection/rapiscan_mp100 (accessed on 26 October 2018).

62. Ziock, K.P.; Hailey, C.J.; Gosnell, T.B.; Lupton, J.H. A Gamma-Ray Imager for Arms Control. *IEEE Trans. Nucl. Sci.* **1992**, *39*, 1046–1050. [CrossRef]

63. Ziock, K.P.; Cheriyadat, A.; Fabris, L.; Goddard, J.; Hornback, D.; Karnowski, T.; Kerekes, R.; Newby, J. Autonomous radiation monitoring of small vessels. *Nucl. Instrum. Methods Phys. Res. Sect. A-Accel. Spectrometers Detect. Assoc. Equip.* **2011**, *652*, 10–15. [CrossRef]

64. Ely, J.; Kouzes, R.; Schweppe, J.; Siciliano, E.; Strachan, D.; Weier, D. The use of energy windowing to discriminate SNM from NORM in radiation portal monitors. *Nucl. Instrum. Methods Phys. Res. Sect. A-Accel. Spectrometers Detect. Assoc. Equip.* **2006**, *560*, 373–387. [CrossRef]

65. Hevener, R.; Yim, M.-S.; Baird, K. Investigation of energy windowing algorithms for effective cargo screening with radiation portal monitors. *Radiat. Meas.* **2013**, *58*, 113–120. [CrossRef]

66. Lo Presti, C.A.; Weier, D.R.; Kouzes, R.T.; Schweppe, J.E. Baseline suppression of vehicle portal monitor gamma count profiles: A characterization study. *Nucl. Instrum. Methods Phys. Res. Sect. A Accel. Spectrometers Detect. Assoc. Equip.* **2006**, *562*, 281–297. [CrossRef]

67. Ziock, K.P.; Goldstein, W.H. The lost source, varying backgrounds and why bigger may not be better. In Proceedings of the Workshop on Unattended Radiation Sensor Systems for Remote Applications, Washington, DC, USA, 15–17 April 2002; pp. 60–70.

68. Ziock, K.P.; Bradley, E.C.; Cheriyadat, A.; Cunningham, M.; Fabris, L.; Fitzgerald, C.L.; Goddard, J.S.; Hornback, D.E.; Kerekes, R.A.; Karnowski, T.P.; et al. Performance of the Roadside Tracker Portal-Less Portal Monitor. *IEEE Trans. Nucl. Sci.* **2013**, *60*, 2237–2246. [CrossRef]

69. Kouzes, R.T.; Siciliano, E.R. The response of radiation portal monitors to medical radionuclides at border crossings. *Radiat. Meas.* **2006**, *41*, 499–512. [CrossRef]

70. Robinson, S.M.; Bender, S.E.; Flumerfelt, E.L.; LoPresti, C.A.; Woodring, M.L. Time Series Evaluation of Radiation Portal Monitor Data for Point Source Detection. *IEEE Trans. Nucl. Sci.* **2009**, *56*, 3688–3693. [CrossRef]

71. Ivanov, O.P.; Semin, I.A.; Potapov, V.N.; Stepanov, V.E. Extra-light gamma-ray imager for safeguards and homeland security. In Proceedings of the 2015 4th International Conference on Advancements in Nuclear Instrumentation Measurement Methods and their Applications, Lisbon, Portugal, 20–24 April 2015.

72. Carrel, F.; Khalil, R.A.; Colas, S.; Toro, D.D.; Ferrand, G.; Gaillard-Lecanu, E.; Gmar, M.; Hameau, D.; Jahan, S.; Lainé, F.; et al. GAMPIX: A new gamma imaging system for radiological safety and Homeland Security Purposes. In Proceedings of the 2011 IEEE Nuclear Science Symposium Conference Record, Valencia, Spain, 23–29 October 2011; pp. 4739–4744.

73. Gal, O.; Izac, C.; Jean, F.; Lainé, F.; Lévêque, C.; Nguyen, A. CARTOGAM—A portable gamma camera for remote localisation of radioactive sources in nuclear facilities. *Nucl. Instrum. Methods Phys. Res. Sect. A Accel. Spectrometers Detect. Assoc. Equip.* **2001**, *460*, 138–145. [CrossRef]

74. Woodring, M.; Souza, D.; Tipnis, S.; Waer, P.; Squillante, M.; Entine, G.; Ziock, K.P. Advanced radiation imaging of low-intensity gamma-ray sources. *Nucl. Instrum. Methods Phys. Res. Sect. A Accel. Spectrometers Detect. Assoc. Equip.* **1999**, *422*, 709–712. [CrossRef]

75. Dubos, S.; Lemaire, H.; Schanne, S.; Limousin, O.; Carrel, F.; Schoepff, V.; Blondel, C. ORIGAMIX, a CdTe-based spectro-imager development for nuclear applications. *Nucl. Instrum. Methods Phys. Res. Sect. A Accel. Spectrometers Detect. Assoc. Equip.* **2015**, *787*, 302–307. [CrossRef]

76. Wulf, E.A.; Phlips, B.F.; Johnson, W.N.; Leas, B.; Mitchell, L.J. MISTI imaging and source localization. In Proceedings of the 2008 IEEE Nuclear Science Symposium Conference Record, Dresden, Germany, 19–25 October 2008; pp. 2413–2417.

77. Kowash, B.R.; Wehe, D.K.; Fessler, J.A. A rotating modulation imager for locating mid-range point sources. *Nucl. Instrum. Methods Phys. Res. Sect. A Accel. Spectrometers Detect. Assoc. Equip.* **2009**, *602*, 477–483. [CrossRef]

78. Vaska, P.; Vanier, P.E.; Junnarkar, S.; Krishnamoorthy, S.; Pratte, J.F.; Stoll, S. A compact scintillator-based coded aperture imager for localizing illicit nuclear materials. In Proceedings of the IEEE Nuclear Science Symposium Conference Record, Honolulu, HI, USA, 26 October–3 November 2007; pp. 1195–1197.

79. Jeong, M.; Van, B.; Wells, B.T.; D'Aries, L.J.; Hammig, M.D. Scalable gamma-ray camera for wide-area search based on silicon photomultipliers array. *Rev. Sci. Instrum.* **2018**, *89*, 033106. [CrossRef] [PubMed]

80. Ziock, K.P. Principles and applications of gamma-ray imaging for arms control. *Nucl. Instrum. Methods Phys. Res. Sect. A Accel. Spectrometers Detect. Assoc. Equip.* **2018**, *878*, 191–199. [CrossRef]

81. Kong, Y.; Brands, H.; Glaser, T.; Herbach, C.; Hoy, L.; Kreuels, M.; Küster, M.; Pausch, G.; Petzoldt, J.; Plettner, C.; et al. A Prototype Compton Camera Array for Localization and Identification of Remote Radiation Sources. *IEEE Trans. Nucl. Sci.* **2013**, *60*, 1066–1071. [CrossRef]

82. Saull, P.R.B.; MacLeod, A.M.L.; Sinclair, L.E.; Drouin, P.L.; Erhardt, L.; Hovgaard, J.; Krupskyy, B.; Ueno, R.; Waller, D.; McCann, A. SCoTSS modular survey spectrometer and Compton imager. In Proceedings of the 2016 IEEE Nuclear Science Symposium, Medical Imaging Conference and Room-Temperature Semiconductor Detector Workshop, NSS/MIC/RTSD, Strasbourg, France, 29 October–6 November 2016.

83. MacLeod, A.M.L.; Boyle, P.J.; Hanna, D.S.; Saull, P.R.B.; Sinclair, L.E.; Seywerd, H.C.J. Development of a Compton imager based on bars of scintillator. *Nucl. Instrum. Methods Phys. Res. Sect. A Accel. Spectrometers Detect. Assoc. Equip.* **2014**, *767*, 397–406. [CrossRef]

84. Vetter, K.; Burks, M.; Cork, C.; Cunningham, M.; Chivers, D.; Hull, E.; Krings, T.; Manini, H.; Mihailescu, L.; Nelson, K.; et al. High-sensitivity Compton imaging with position-sensitive Si and Ge detectors. *Nucl. Instrum. Methods Phys. Res. Sect. A Accel. Spectrometers Detect. Assoc. Equip.* **2007**, *579*, 363–366. [CrossRef]

85. Hynes, M.V.; Harris, B.; Lednum, E.E.; Wallace, M.S.; Schultz, L.J.; Palmer, D.M.; Wakeford, D.T.; Andrews, H.R.; Lanza, R.C.; Clifford, E.T.; et al. Multimodal Radiation Imager. U.S. Patent 7,863,567, 4 January 2011.

86. Andreyev, A.; Sitek, A.; Celler, A. Fast image reconstruction for Compton camera using stochastic origin ensemble approach. *Med. Phys.* **2011**, *38*, 429–438. [CrossRef] [PubMed]

87. Hoover, A.S.; Kippen, R.M.; Sullivan, J.P.; Rawool-Sullivan, M.W.; Baird, W.; Sorensen, E.B. The LANL prototype Compton gamma-ray imager: Design and image reconstruction techniques. *IEEE Trans. Nucl. Sci.* **2005**, *52*, 3047–3053. [CrossRef]

88. Montémont, G.; Bohuslav, P.; Dubosq, J.; Feret, B.; Monnet, O.; Oehling, O.; Skala, L.; Stanchina, S.; Verger, L.; Werthmann, G. NuVISION: A Portable Multimode Gamma Camera based on HiSPECT Imaging Module. In Proceedings of the 2017 IEEE Nuclear Science Symposium and Medical Imaging Conference (NSS/MIC), Atlanta, GA, USA, 21–28 October 2017; pp. 1–3.

89. Penny, R.D.; Hood, W.E.; Polichar, R.M.; Cardone, F.H.; Chavez, L.G.; Grubbs, S.G.; Huntley, B.P.; Kuharski, R.A.; Shyffer, R.T.; Fabris, L.; et al. A dual-sided coded-aperture radiation detection system. *Nucl. Instrum. Methods Phys. Res. Sect. A Accel. Spectrometers Detect. Assoc. Equip.* **2011**, *652*, 578–581. [CrossRef]

90. Zelakiewicz, S.; Hoctor, R.; Ivan, A.; Ross, W.; Nieters, E.; Smith, W.; McDevitt, D.; Wittbrodt, M.; Milbrath, B. SORIS-A standoff radiation imaging system. *Nucl. Instrum. Methods Phys. Res. Sect. A Accel. Spectrometers Detect. Assoc. Equip.* **2011**, *652*, 5–9. [CrossRef]

91. Boehnen, C.; Paquit, V.; Ziock, K.; Guzzardo, T.; Whitaker, M.; Raffo-Caiado, A. Field trial of a highly portable coded aperture gamma ray and 3D imaging system. In Proceedings of the 2011 Future of Instrumentation International Workshop (FIIW), Oak Ridge, TN, USA, 7–8 November 2011; pp. 75–78.

92. Tornga, S.R.; Sullivan, M.W.R.; Sullivan, J.P. Three-Dimensional Compton Imaging Using List-Mode Maximum Likelihood Expectation Maximization. *IEEE Trans. Nucl. Sci.* **2009**, *56*, 1372–1376. [CrossRef]

93. Kouzes, R.T.; Ely, J.H.; Erikson, L.E.; Kernan, W.J.; Lintereur, A.T.; Siciliano, E.R.; Stephens, D.L.; Stromswold, D.C.; Van Ginhoven, R.M.; Woodring, M.L. Neutron detection alternatives to ^{3}HE for national security applications. *Nucl. Instrum. Methods Phys. Res. Sect. A Accel. Spectrometers Detect. Assoc. Equip.* **2010**, *623*, 1035–1045. [CrossRef]

94. Peerani, P.; Tomanin, A.; Pozzi, S.; Dolan, J.; Miller, E.; Flaska, M.; Battaglieri, M.; De Vita, R.; Ficini, L.; Ottonello, G.; et al. Testing on novel neutron detectors as alternative to ^{3}He for security applications. *Nucl. Instrum. Methods Phys. Res. Sect. A Accel. Spectrometers Detect. Assoc. Equip.* **2012**, *696*, 110–120. [CrossRef]

95. Lintereur, A.T.; Ely, J.H.; Kouzes, R.T.; Siciliano, E.R.; Swinhoe, M.T.; Woodring, M.L. Alternatives to Helium-3 for Neutron Multiplicity Counters. In Proceedings of the 2012 IEEE Nuclear Science Symposium and Medical Imaging Conference Record, Anaheim, CA, USA, 27 October–3 November 2012; Yu, B., Ed.; 2012; pp. 547–553.

96. Kouzes, R.T.; Lintereur, A.T.; Siciliano, E.R. Progress in alternative neutron detection to address the helium-3 shortage. *Nucl. Instrum. Methods Phys. Res. Sect. A Accel. Spectrometers Detect. Assoc. Equip.* **2015**, *784*, 172–175. [CrossRef]

97. Zeitelhack, K. Search for alternative techniques to helium-3 based detectors for neutron scattering applications. *Neutron News* **2012**, *23*, 10–13. [CrossRef]

98. Lintereur, A.; Conlin, K.; Ely, J.; Erikson, L.; Kouzes, R.; Siciliano, E.; Stromswold, D.; Woodring, M. ^{3}He and BF$_3$ neutron detector pressure effect and model comparison. *Nucl. Instrum. Methods Phys. Res. Sect. A Accel. Spectrometers Detect. Assoc. Equip.* **2011**, *652*, 347–350. [CrossRef]

99. Flaska, M.; Pozzi, S.A.; Czirr, J.B. Use of an LGB detector in nuclear nonproliferation applications. In Proceedings of the 2008 IEEE Nuclear Science Symposium Conference Record, Dresden, Germany, 19–25 October 2008; pp. 3376–3380.

100. Tomanin, A.; Paepen, J.; Schillebeeckx, P.; Wynants, R.; Nolte, R.; Lavietes, A. Characterization of a cubic EJ-309 liquid scintillator detector. *Nucl. Instrum. Methods Phys. Res. Sect. A Accel. Spectrometers Detect. Assoc. Equip.* **2014**, *756*, 45–54. [CrossRef]

101. Pawełczak, I.A.; Glenn, A.M.; Martinez, H.P.; Carman, M.L.; Zaitseva, N.P.; Payne, S.A. Boron-loaded plastic scintillator with neutron-γ pulse shape discrimination capability. *Nucl. Instrum. Methods Phys. Res. Sect. A Accel. Spectrometers Detect. Assoc. Equip.* **2014**, *751*, 62–69. [CrossRef]

102. Birch, J.; Buffet, J.C.; Correa, J.; Esch, P.V.; Guérard, B.; Hall-Wilton, R.; Höglund, C.; Hultman, L.; Khaplanov, A.; Piscitelli, F. (B4C)-B-10 Multi-Grid as an Alternative to He-3 for Large Area Neutron Detectors. *IEEE Trans. Nucl. Sci.* **2013**, *60*, 871–878. [CrossRef]

103. Swiderski, L.; Moszynski, M.; Wolski, D.; Batsch, T.; Nassalski, A.; Syntfeld-Kazuch, A.; Szczesniak, T.; Kniest, F.; Kusner, M.R.; Pausch, G.; et al. Boron-10 Loaded BC523A Liquid Scintillator for Neutron Detection in the Border Monitoring. *IEEE Trans. Nucl. Sci.* **2008**, *55*, 3710–3716. [CrossRef]

104. Lawrence, C.C.; Febbraro, M.; Massey, T.N.; Flaska, M.; Becchetti, F.D.; Pozzi, S.A. Neutron response characterization for an EJ299-33 plastic scintillation detector. *Nucl. Instrum. Methods Phys. Res. Sect. A Accel. Spectrometers Detect. Assoc. Equip.* **2014**, *759*, 16–22. [CrossRef]

105. Mayer, M.; Nattress, J.; Trivelpiece, C.; Jovanovic, I. Geometric optimization of a neutron detector based on a lithium glass–polymer composite. *Nucl. Instrum. Methods Phys. Res. Sect. A Accel. Spectrometers Detect. Assoc. Equip.* **2015**, *784*, 168–171. [CrossRef]

106. Ryzhikov, V.; Nagornaya, L.; Burachas, S.; Piven, L.; Danshin, E.; Zelenskaya, O.; Chernikov, V. Detection of thermal and resonance neutrons using oxide scintillators. *IEEE Trans. Nucl. Sci.* **2000**, *47*, 2061–2064. [CrossRef]

107. Haas, D.A.; Bliss, M.; Bowyer, S.M.; Kephart, J.D.; Schweiger, M.J.; Smith, L.E. Actinide-loaded glass scintillators for fast neutron detection. *Nucl. Instrum. Methods Phys. Res. Sect. A Accel. Spectrometers Detect. Assoc. Equip.* **2011**, *652*, 421–423. [CrossRef]

108. Van Eijk, C.W.E. Inorganic Scintillators for Thermal Neutron Detection. *IEEE Trans. Nucl. Sci.* **2012**, *59*, 2242–2247. [CrossRef]

109. Seymour, R.S.; Richardson, B.; Morichi, M.; Bliss, M.; Craig, R.A.; Sunberg, D.S. Scintillating-Glass-Fiber Neutron Sensors, their Application and Performance for Plutonium Detection and Monitoring. *J. Radioanal. Nucl. Chem.* **2000**, *243*, 387–388. [CrossRef]

110. Jordan, D.V.; Ely, J.H.; Peurrung, A.J.; Bond, L.J.; Collar, J.I.; Flake, M.; Knopf, M.A.; Pitts, W.K.; Shaver, M.; Sonnenschein, A.; et al. Neutron detection via bubble chambers. *Appl. Radiat. Isot.* **2005**, *63*, 645–653. [CrossRef] [PubMed]

111. Bramblett, R.L.; Ewing, R.I.; Bonner, T.W. A new type of neutron spectrometer. *Nucl. Instrum. Methods* **1960**, *9*, 1–12. [CrossRef]

112. Thomas, D.J.; Alevra, A.V. Bonner sphere spectrometers—A critical review. *Nucl. Instrum. Methods Phys. Res. Sect. A Accel. Spectrometers Detect. Assoc. Equip.* **2002**, *476*, 12–20. [CrossRef]

113. Caruso, A.N. The physics of solid-state neutron detector materials and geometries. *J. Phys. Condens. Matter* **2010**, *22*, 443201. [CrossRef] [PubMed]

114. Peurrung, A.J. Recent developments in neutron detection. *Nucl. Instrum. Methods Phys. Res. Sect. A Accel. Spectrometers Detect. Assoc. Equip.* **2000**, *443*, 400–415. [CrossRef]

115. Runkle, R.C.; Bernstein, A.; Vanier, P.E. Securing special nuclear material: Recent advances in neutron detection and their role in nonproliferation. *J. Appl. Phys.* **2010**, *108*, 13. [CrossRef]

116. Szalkai, D.; Ferone, R.; Gehre, D.; Issa, F.; Klix, A.; Lyoussi, A.; Ottaviani, L.; Rücker, T.; Tüttő, P.; Vervisch, V. Detection of 14 MeV neutrons in high temperature environment up to 500 °C using 4H-SiC based diode detector. In Proceedings of the 2015 4th International Conference on Advancements in Nuclear Instrumentation Measurement Methods and their Applications (ANIMMA), Lisbon, Portugal, 20–24 April 2015; pp. 1–6.

117. Ha, J.H.; Kang, S.M.; Park, S.H.; Kim, H.S.; Lee, N.H.; Song, T.-Y. A self-biased neutron detector based on an SiC semiconductor for a harsh environment. *Appl. Radiat. Isot.* **2009**, *67*, 1204–1207. [CrossRef] [PubMed]

118. Balmer, R.S.; Brandon, J.R.; Clewes, S.L.; Dhillon, H.K.; Dodson, J.M.; Friel, I.; Inglis, P.N.; Madgwick, T.D.; Markham, M.L.; Mollart, T.P.; et al. Chemical vapour deposition synthetic diamond: Materials, technology and applications. *J. Phys. Condens. Matter* **2009**, *21*, 364221. [CrossRef] [PubMed]

119. Dumazert, J.; Coulon, R.; Kondrasovs, V.; Boudergui, K. Compensation scheme for online neutron detection using a Gd-covered CdZnTe sensor. *Nucl. Instrum. Methods Phys. Res. Sect. A-Accel. Spectrometers Detect. Assoc. Equip.* **2017**, *857*, 7–15. [CrossRef]

120. Obraztsova, O.; Ottaviani, L.; Klix, A.; Döring, T.; Palais, O.; Lyoussi, A. Comparison between Silicon-Carbide and diamond for fast neutron detection at room temperature. *EPJ Web Conf.* **2018**, *170*. [CrossRef]

121. Streicher, M.; Goodman, D.; Zhu, Y.; Brown, S.; Kiff, S.; He, Z. Fast Neutron Detection Using Pixelated CdZnTe Spectrometers. *IEEE Trans. Nucl. Sci.* **2017**, *64*, 1920–1926. [CrossRef]

122. Janssens-Maenhout, G.; De Roo, F.; Janssens, W. Contributing to shipping container security: Can passive sensors bring a solution? *J. Environ. Radioact.* **2010**, *101*, 95–105. [CrossRef]

123. Kouzes, R.T.; Ely, J.H.; Lintereur, A.T.; Mace, E.K.; Stephens, D.L.; Woodring, M.L. Neutron detection gamma ray sensitivity criteria. *Nucl. Instrum. Methods Phys. Res. Sect. A Accel. Spectrometers Detect. Assoc. Equip.* **2011**, *654*, 412–416. [CrossRef]

124. Stave, S.; Bliss, M.; Kouzes, R.; Lintereur, A.; Robinson, S.; Siciliano, E.; Wood, L. LiF/ZnS neutron multiplicity counter. *Nucl. Instrum. Methods Phys. Res. Sect. A Accel. Spectrometers Detect. Assoc. Equip.* **2015**, *784*, 208–212. [CrossRef]

125. Oakes, T.M.; Bellinger, S.L.; Miller, W.H.; Myers, E.R.; Fronk, R.G.; Cooper, B.W.; Sobering, T.J.; Scott, P.R.; Ugorowski, P.; McGregor, D.S.; et al. An accurate and portable solid state neutron rem meter. *Nucl. Instrum. Methods Phys. Res. Sect. A Accel. Spectrometers Detect. Assoc. Equip.* **2013**, *719*, 6–12. [CrossRef]

126. Kouzes, R.T.; Ely, J.H.; Lintereur, A.T.; Siciliano, E.R. Boron-10 based neutron coincidence counter for safeguards. *IEEE Trans. Nucl. Sci.* **2014**, *61*, 2608–2618. [CrossRef]

127. Littell, J.; Lukosi, E.; Hayward, J.; Milburn, R.; Rowan, A. Coded moderator approach for fast neutron source detection and localization at standoff. *Nucl. Instrum. Methods Phys. Res. Sect. A Accel. Spectrometers Detect. Assoc. Equip.* **2015**, *784*, 364–369. [CrossRef]

128. Runkle, R.C. Neutron sensors and their role in nuclear nonproliferation. *Nucl. Instrum. Methods Phys. Res. Sect. A Accel. Spectrometers Detect. Assoc. Equip.* **2011**, *652*, 37–40. [CrossRef]

129. Gamage, K.A.A.; Joyce, M.J.; Adams, J.C. Combined digital imaging of mixed-field radioactivity with a single detector. *Nucl. Instrum. Methods Phys. Res. Sect. A Accel. Spectrometers Detect. Assoc. Equip.* **2011**, *635*, 74–77. [CrossRef]

130. Flaska, M.; Pozzi, S.A. Digital pulse shape analysis for the capture-gated liquid scintillator BC-523A. *Nucl. Instrum. Methods Phys. Res. Sect. A-Accel. Spectrometers Detect. Assoc. Equip.* **2009**, *599*, 221–225. [CrossRef]

131. Pausch, G.; Stein, J. Application of6LiI(Eu) scintillators with photodiode readout for neutron counting in mixed gamma-neutron fields. *IEEE Trans. Nucl. Sci.* **2008**, *55*, 1413–1419. [CrossRef]

132. Joyce, M.J.; Gamage, K.A.A. Real-time, digital imaging of fast neutrons and γ rays with a single fast liquid scintillation detector. In Proceedings of the IEEE Nuclear Science Symposium Conference Record, Honolulu, HI, USA, 26 October–3 November 2007; pp. 602–606.

133. Payne, C.; Sellin, P.J.; Ellis, M.; Duroe, K.; Jones, A.; Joyce, M.; Randall, G.; Speller, R. Neutron/gamma pulse shape discrimination in EJ-299-34 at high flux. In Proceedings of the 2015 IEEE Nuclear Science Symposium and Medical Imaging Conference, NSS/MIC, San Diego, CA, USA, 31 October–7 November 2015.

134. Liu, G.; Joyce, M.J.; Ma, X.; Aspinall, M.D. A digital method for the discrimination of neutrons and γ rays with organic scintillation detectors using frequency gradient analysis. *IEEE Trans. Nucl. Sci.* **2010**, *57*, 1682–1691. [CrossRef]

135. Unsurpassed Mobile Primary Screening. Available online: http://www.symetrica.com/mobile-rpm (accessed on 28 January 2019).

136. Fission Meter Portable Neutron Source Identification System. Available online: https://www.ortec-online.com/products/nuclear-security-and-safeguards/neutron-fission-systems/fission-meter (accessed on 28 January 2019).

137. Smiths Detection Radseeker. Available online: http://www.symetrica.com/oem-sub-systems (accessed on 28 January 2019).

138. Flat Panel Backpack Neutron Detection (^{3}HE Free). Available online: http://www.symetrica.com/backpack (accessed on 26 October 2018).

139. Miller, R.S.; Macri, J.R.; McConnell, M.L.; Ryan, J.M.; Flückiger, E.; Desorgher, L. SONTRAC: An imaging spectrometer for MeV neutrons. *Nucl. Instrum. Methods Phys. Res. Sect. A Accel. Spectrometers Detect. Assoc. Equip.* **2003**, *505*, 36–40. [CrossRef]

140. Bravar, U.; Bruillard, P.J.; Flckiger, E.O.; Macri, J.R.; McConnell, M.L.; Moser, M.R.; Ryan, J.M.; Woolf, R.S. Design and Testing of a Position-Sensitive Plastic Scintillator Detector for Fast Neutron Imaging. *IEEE Trans. Nucl. Sci.* **2006**, *53*, 3894–3903. [CrossRef]

141. Vanier, P.E.; Forman, L.; Dioszegi, I.; Salwen, C.; Ghosh, V.J. Calibration and testing of a large-area fast-neutron directional detector. In Proceedings of the 2007 IEEE Nuclear Science Symposium Conference Record, Honolulu, HI, USA, 26 October–3 November 2007; pp. 179–184.

142. Mascarenhas, N.; Brennan, J.; Krenz, K.; Marleau, P.; Mrowka, S. Results with the Neutron Scatter Camera. *IEEE Trans. Nucl. Sci.* **2009**, *56*, 1269–1273. [CrossRef]

143. Siegmund, O.H.W.; Vallerga, J.V.; Tremsin, A.S.; Feller, W.B. High spatial and temporal resolution neutron imaging with microchannel plate detectors. *IEEE Trans. Nucl. Sci.* **2009**, *56*, 1203–1209. [CrossRef]

144. Herbach, C.; Pausch, G.; Kreuels, A.; Kong, Y.; Lentering, R.; Plettner, C.; Roemer, K.; Scherwinski, F.; Schotanus, P.; Stein, J.; et al. Neutron detection by measuring capture gammas in a calorimetric approach. In Proceedings of the IEEE Nuclear Science Symposuim & Medical Imaging Conference, Knoxville, TN, USA, 30 October–6 November 2010; pp. 1827–1834.

145. Ryzhikov, V.D.; Grinyov, B.V.; Onyshchenko, G.M.; Piven, L.A.; Lysetska, O.K.; Nagornaya, L.L.; Pochet, T. The Use of Fast and Thermal Neutron Detectors Based on Oxide Scintillators in Inspection Systems for Prevention of Illegal Transportation of Radioactive Substances. *IEEE Trans. Nucl. Sci.* **2010**, *57*, 2747–2751. [CrossRef]

146. Marleau, P.; Brennan, J.; Brubaker, E.; Steele, J. Results from the Coded Aperture Neutron Imaging System. In Proceedings of the 2010 IEEE Nuclear Science Symposium Conference Record, Knoxville, TN, USA, 30 October–6 November 2010; pp. 1640–1646.

147. Nakae, L.F.; Chapline, G.F.; Glenn, A.M.; Kerr, P.L.; Kim, K.S.; Ouedraogo, S.A.; Prasad, M.K.; Sheets, S.A.; Snyderman, N.J.; Verbeke, J.M.; et al. Recent developments in fast neutron detection and multiplicity counting with liquid scintillator. *AIP Conf. Proc.* **2011**, *1412*, 240–248.

148. Bellinger, S.L.; Fronk, R.G.; Sobering, T.J.; McGregor, D.S. High-efficiency microstructured semiconductor neutron detectors that are arrayed, dual-integrated, and stacked. *Appl. Radiat. Isot.* **2012**, *70*, 1121–1124. [CrossRef] [PubMed]

149. Ide, K.; Becchetti, M.F.; Flaska, M.; Poitrasson-Riviere, A.; Hamel, M.C.; Polack, J.K.; Lawrence, C.C.; Clarke, S.D.; Pozzi, S.A. Analysis of a measured neutron background below 6MeV for fast-neutron imaging systems. *Nucl. Instrum. Methods Phys. Res. Sect. A Accel. Spectrometers Detect. Assoc. Equip.* **2012**, *694*, 24–31. [CrossRef]

150. Joyce, M.J.; Gamage, K.A.A.; Aspinall, M.D.; Cave, F.D.; Lavietes, A. Real-Time, Fast Neutron Coincidence Assay of Plutonium With a 4-Channel Multiplexed Analyzer and Organic Scintillators. *IEEE Trans. Nucl. Sci.* **2014**, *61*, 1340–1348. [CrossRef]

151. Brennan, J.; Brubaker, E.; Gerling, M.; Marleau, P.; McMillan, K.; Nowack, A.; Galloudec, N.R.-L.; Sweany, M. Demonstration of two-dimensional time-encoded imaging of fast neutrons. *Nucl. Instrum. Methods Phys. Res. Sect. A Accel. Spectrometers Detect. Assoc. Equip.* **2015**, *802*, 76–81. [CrossRef]

152. Fronk, R.G.; Bellinger, S.L.; Henson, L.C.; Huddleston, D.E.; Ochs, T.R.; Rietcheck, C.J.; Smith, C.T.; Shultis, J.K.; Sobering, T.J.; McGregor, D.S. Advancements on dual-sided microstructured semiconductor neutron detectors (DSMSNDs). In Proceedings of the 2015 IEEE Nuclear Science Symposium and Medical Imaging Conference (NSS/MIC), San Diego, CA, USA, 31 October–7 November 2015; pp. 1–4.

153. Ianakiev, K.D.; Hehlen, M.P.; Swinhoe, M.T.; Favalli, A.; Iliev, M.L.; Lin, T.C.; Bennett, B.L.; Barker, M.T. Neutron detector based on Particles of 6Li glass scintillator dispersed in organic lightguide matrix. *Nucl. Instrum. Methods Phys. Res. Sect. A Accel. Spectrometers Detect. Assoc. Equip.* **2015**, *784*, 189–193. [CrossRef]

154. Hoshor, C.B.; Oakes, T.M.; Myers, E.R.; Rogers, B.J.; Currie, J.E.; Young, S.M.; Crow, J.A.; Scott, P.R.; Miller, W.H.; Bellinger, S.L.; et al. A portable and wide energy range semiconductor-based neutron spectrometer. *Nucl. Instrum. Methods Phys. Res. Sect. A Accel. Spectrometers Detect. Assoc. Equip.* **2015**, *803*, 68–81. [CrossRef]

155. Goldsmith, J.E.M.; Gerling, M.D.; Brennan, J.S. A compact neutron scatter camera for field deployment. *Rev. Sci. Instrum.* **2016**, *87*, 083307. [CrossRef] [PubMed]

156. Di Fulvio, A.; Shin, T.H.; Jordan, T.; Sosa, C.; Ruch, M.L.; Clarke, S.D.; Chichester, D.L.; Pozzi, S.A. Passive assay of plutonium metal plates using a fast-neutron multiplicity counter. *Nucl. Instrum. Methods Phys. Res. Sect. A Accel. Spectrometers Detect. Assoc. Equip.* **2017**, *855*, 92–101. [CrossRef]

157. Cowles, C.; Behling, S.; Baldez, P.; Folsom, M.; Kouzes, R.; Kukharev, V.; Lintereur, A.; Robinson, S.; Siciliano, E.; Stave, S.; et al. Development of a lithium fluoride zinc sulfide based neutron multiplicity counter. *Nucl. Instrum. Methods Phys. Res. Sect. A Accel. Spectrometers Detect. Assoc. Equip.* **2018**, *887*, 59–63. [CrossRef]

158. Ochs, T.R.; Beatty, B.L.; Bellinger, S.L.; Fronk, R.G.; Gardner, J.A.; Henson, L.C.; Huddleston, D.E.; Hutchins, R.M.; Sobering, T.J.; Thompson, J.L.; et al. Wearable detector device utilizing microstructured semiconductor neutron detector technology. *Radiat. Phys. Chem.* **2019**, *155*, 164–172. [CrossRef]

159. Brooks, F.D. A scintillation counter with neutron and gamma-ray discriminators. *Nucl. Instrum. Methods* **1959**, *4*, 151–163. [CrossRef]

160. Adams, J.M.; White, G. A versatile pulse shape discriminator for charged particle separation and its application to fast neutron time-of-flight spectroscopy. *Nucl. Instrum. Methods* **1978**, *156*, 459–476. [CrossRef]

161. Yang, K.; Menge, P.R.; Ouspenski, V. Li Co-Doped NaI:Tl (NaIL)-A Large Volume Neutron-Gamma Scintillator with Exceptional Pulse Shape Discrimination. *IEEE Trans. Nucl. Sci.* **2017**, *64*, 2406–2413. [CrossRef]

162. Mukhopadhyay, S.; McHugh, H.R. Portable gamma and thermal neutron detector using 6LiI(Eu) crystals. In Proceedings of the Proceedings of SPIE-The International Society for Optical Engineering, San Diego, CA, USA, 20 January 2004; pp. 73–82.

163. Soundara-Pandian, L.; Hawrami, R.; Glodo, J.; Ariesanti, E.; Loef, E.V.; Shah, K. Lithium Alkaline Halides—Next Generation of Dual Mode Scintillators. *IEEE Trans. Nucl. Sci.* **2016**, *63*, 490–496. [CrossRef]

164. McGregor, D.S.; Lindsay, J.T.; Olsen, R.W. Thermal neutron detection with cadmium1-x zincx telluride semiconductor detectors. *Nucl. Instrum. Methods Phys. Res. Sect. A-Accel. Spectrometers Detect. Assoc. Equip.* **1996**, *381*, 498–501. [CrossRef]

165. Martín-Martín, A.; Iñiguez, M.P.; Luke, P.N.; Barquero, R.; Lorente, A.; Morchón, J.; Gallego, E.; Quincoces, G.; Martí-Climent, J.M. Evaluation of CdZnTe as neutron detector around medical accelerators. *Radiat. Prot. Dosim.* **2009**, *133*, 193–199. [CrossRef]

166. Tupitsyn, E.; Bhattacharya, P.; Rowe, E.; Matei, L.; Groza, M.; Wiggins, B.; Burger, A.; Stowe, A. Single crystal of LiInSe2 semiconductor for neutron detector. *Appl. Phys. Lett.* **2012**, *101*, 202101. [CrossRef]

167. Coceva, C. Pulse-shape discrimination with a glass scintillator. *Nucl. Instrum. Methods* **1963**, *21*, 93–96. [CrossRef]

168. Combes, C.M.; Dorenbos, P.; van Eijk, C.W.E.; Kramer, K.W.; Gudel, H.U. Optical and scintillation properties of pure and Ce^{3+}-doped Cs_2LiYCl_6 and Li_3YCl_6: Ce^{3+} crystals. *J. Lumin.* **1999**, *82*, 299–305. [CrossRef]

169. Glodo, J.; Wang, Y.; Shawgo, R.; Brecher, C.; Hawrami, R.H.; Tower, J.; Shah, K.S. New Developments in Scintillators for Security Applications. *Phys. Procedia* **2017**, *90*, 285–290. [CrossRef]

170. Bell, Z.W. Tests on a digital neutron-gamma pulse shape discriminator with NE213. *Nucl. Instrum. Methods Phys. Res.* **1981**, *188*, 105–109. [CrossRef]

171. Kaschuck, Y.; Esposito, B. Neutron/γ-ray digital pulse shape discrimination with organic scintillators. *Nucl. Instrum. Methods Phys. Res. Sect. A Accel. Spectrometers Detect. Assoc. Equip.* **2005**, *551*, 420–428. [CrossRef]

172. Pozzi, S.A.; Bourne, M.M.; Clarke, S.D. Pulse shape discrimination in the plastic scintillator EJ-299-33. *Nucl. Instrum. Methods Phys. Res. Sect. A-Accel. Spectrometers Detect. Assoc. Equip.* **2013**, *723*, 19–23. [CrossRef]

173. Stevanato, L.; Cester, D.; Nebbia, G.; Viesti, G. Neutron detection in a high gamma-ray background with EJ-301 and EJ-309 liquid scintillators. *Nucl. Instrum. Methods Phys. Res. Sect. A Accel. Spectrometers Detect. Assoc. Equip.* **2012**, *690*, 96–101. [CrossRef]

174. Al Hamrashdi, H.; Cheneler, D.; Monk, S.D. Material optimization in dual particle detectors by comparing advanced scintillating materials using two Monte Carlo codes. *Nucl. Instrum. Methods Phys. Res. Sect. A Accel. Spectrometers Detect. Assoc. Equip.* **2017**, *869*, 163–171. [CrossRef]

175. Bell, Z.W.; Hornback, D.E.; Hu, M.Z.; Neal, J.S. Wavelength-based neutron/gamma ray discrimination in CLYC. In Proceedings of the 2014 IEEE Nuclear Science Symposium and Medical Imaging Conference (NSS/MIC), Seattle, WA, USA, 8–15 November 2014; pp. 1–8.

176. Cester, D.; Nebbia, G.; Stevanato, L.; Pino, F.; Sajo-Bohus, L.; Viesti, G. A compact neutron–gamma spectrometer. *Nucl. Instrum. Methods Phys. Res. Sect. A Accel. Spectrometers Detect. Assoc. Equip.* **2013**, *719*, 81–84. [CrossRef]

177. Paff, M.G.; Ruch, M.L.; Poitrasson-Riviere, A.; Sagadevan, A.; Clarke, S.D.; Pozzi, S. Organic liquid scintillation detectors for on-the-fly neutron/gamma alarming and radionuclide identification in a pedestrian radiation portal monitor. *Nucl. Instrum. Methods Phys. Res. Sect. A Accel. Spectrometers Detect. Assoc. Equip.* **2015**, *789*, 16–27. [CrossRef]

178. Soundara-Pandian, L.; Tower, J.; Hines, C.; O'Dougherty, P.; Glodo, J.; Shah, K. Characterization of Large Volume CLYC Scintillators for Nuclear Security Applications. *IEEE Trans. Nucl. Sci.* **2017**, *64*, 1744–1748. [CrossRef]

179. Gamage, K.A.A.; Joyce, M.J.; Taylor, G.C. A digital approach to neutron–γ imaging with a narrow tungsten collimator aperture and a fast organic liquid scintillator detector. *Appl. Radiat. Isot.* **2012**, *70*, 1223–1227. [CrossRef] [PubMed]

180. Soundara-Pandian, L.; Whitney, C.; Christian, J.; Glodo, J.; Gueorgiev, A.; Hawrami, R.; Squillante, M.R.; Shah, K.S. CLYC in gamma -Neutron imaging system. In Proceedings of the IEEE Nuclear Science Symposium Conference Record, Honolulu, HI, USA, 26 October–3 November 2007; pp. 101–105.

181. McDonald, B.S.; Myjak, M.J.; Zalavadia, M.A.; Smart, J.E.; Willett, J.A.; Landgren, P.C.; Greulich, C.R. A wearable sensor based on CLYC scintillators. *Nucl. Instrum. Methods Phys. Res. Sect. A Accel. Spectrometers Detect. Assoc. Equip.* **2016**, *821*, 73–80. [CrossRef]

182. Budden, B.S.; Stonehill, L.C.; Dallmann, N.; Baginski, M.J.; Best, D.J.; Smith, M.B.; Graham, S.A.; Dathy, C.; Frank, J.M.; McClish, M. A Cs2LiYCl6: Ce-based advanced radiation monitoring device. *Nucl. Instrum. Methods Phys. Res. Sect. A Accel. Spectrometers Detect. Assoc. Equip.* **2015**, *784*, 97–104. [CrossRef]

183. Glodo, J.; Brys, W.; Entine, G.; Higgins, W.M.; Loef, E.V.D.v.; Squillante, M.R.; Shah, K.S. CS2LiYCl6: Ce Neutron gamma detection system. In Proceedings of the 2007 IEEE Nuclear Science Symposium Conference Record, Honolulu, HI, USA, 26 October–3 November 2007; pp. 959–962.

184. Aryaeinejad, R.; Reber, E.L.; Spencer, D.F. Development of a handheld device for simultaneous monitoring of fast neutrons and gamma rays. *IEEE Trans. Nucl. Sci.* **2002**, *49*, 1909–1913. [CrossRef]

185. Aryaeinejad, R.; Spencer, D.F. Pocket dual neutron/gamma radiation detector. *IEEE Trans. Nucl. Sci.* **2004**, *51*, 1667–1671. [CrossRef]

186. Baker, J.H.; Galunov, N.Z.; Seminozhenko, V.P.; Tarasenko, O.A.; Martynenko, E.V. A combined NaI(Tl)+LiI(Eu) detector for environmental, geological and security applications. *Radiat. Meas.* **2007**, *42*, 937–940. [CrossRef]

187. Enqvist, A.; Flaska, M.; Pozzi, S. Measurement and simulation of neutron/gamma-ray cross-correlation functions from spontaneous fission. *Nucl. Instrum. Methods Phys. Res. Sect. A Accel. Spectrometers Detect. Assoc. Equip.* **2008**, *595*, 426–430. [CrossRef]

188. Runkle, R.C.; Myjak, M.J.; Kiff, S.D.; Sidor, D.E.; Morris, S.J.; Rohrer, J.S.; Jarman, K.D.; Pfund, D.M.; Todd, L.C.; Bowler, R.S.; et al. Lynx: An unattended sensor system for detection of gamma-ray and neutron emissions from special nuclear materials. *Nucl. Instrum. Methods Phys. Res. Sect. A Accel. Spectrometers Detect. Assoc. Equip.* **2009**, *598*, 815–825. [CrossRef]

189. Polack, J.K.; Poitrasson-Rivière, A.; Hamel, M.C.; Ide, K.; McMillan, K.L.; Clarke, S.D.; Flaska, M.; Pozzi, S.A. Dual-particle imager for standoff detection of special nuclear material. In Proceedings of the 2011 IEEE Nuclear Science Symposium Conference Record, Valencia, Spain, 23–29 October 2011; pp. 1494–1500.

190. Cester, D.; Nebbia, G.; Stevanato, L.; Viesti, G.; Neri, F.; Petrucci, S.; Selmi, S.; Tintori, C.; Peerani, P.; Tomanin, A. Special nuclear material detection with a mobile multi-detector system. *Nucl. Instrum. Methods Phys. Res. Sect. A Accel. Spectrometers Detect. Assoc. Equip.* **2012**, *663*, 55–63. [CrossRef]

191. Ayaz-Maierhafer, B.; Hayward, J.P.; Ziock, K.P.; Blackston, M.A.; Fabris, L. Angular resolution study of a combined gamma-neutron coded aperture imager for standoff detection. *Nucl. Instrum. Methods Phys. Res. Sect. A Accel. Spectrometers Detect. Assoc. Equip.* **2013**, *712*, 120–125. [CrossRef]

192. Poitrasson-Riviere, A.; Hamel, M.C.; Polack, J.K.; Flaska, M.; Clarke, S.D.; Pozzi, S.A. Dual-particle imaging system based on simultaneous detection of photon and neutron collision events. *Nucl. Instrum. Methods Phys. Res. Sect. A Accel. Spectrometers Detect. Assoc. Equip.* **2014**, *760*, 40–45. [CrossRef]

193. Cester, D.; Lunardon, M.; Moretto, S.; Nebbia, G.; Pino, F.; Sajo-Bohus, L.; Stevanato, L.; Bonesso, I.; Turato, F. A novel detector assembly for detecting thermal neutrons, fast neutrons and gamma rays. *Nucl. Instrum. Methods Phys. Res. Sect. A Accel. Spectrometers Detect. Assoc. Equip.* **2016**, *830*, 191–196. [CrossRef]

194. Nemzek, R.; Kenyon, G.; Koehler, A.; Lee, D.M.; Priedhorsky, W.; Raby, E.Y. SNM-DAT: Simulation of a heterogeneous network for nuclear border security. *Nucl. Instrum. Methods Phys. Res. Sect. A Accel. Spectrometers Detect. Assoc. Equip.* **2007**, *579*, 414–417. [CrossRef]

195. Cooper, D.A.; Ledoux, R.J.; Kamieniecki, K.; Korbly, S.E.; Thompson, J.; Ryan, M.; Roza, N.; Perry, L.; Hwang, D.; Costales, J.; et al. Intelligent radiation sensor system (IRSS) advanced technology demonstration (ATD). In Proceedings of the 2010 IEEE International Conference on Technologies for Homeland Security, HST, Waltham, MA, USA, 13–15 November 2010; pp. 414–420.

196. Vilim, R.; Klann, R. RadTrac: A System for Detecting, Localizing, and Tracking Radioactive Sources in Real Time. *Nucl. Technol.* **2009**, *168*, 61–73. [CrossRef]

197. Rao, N.S.V.; Sen, S.; Prins, N.J.; Cooper, D.A.; Ledoux, R.J.; Costales, J.B.; Kamieniecki, K.; Korbly, S.E.; Thompson, J.K.; Batcheler, J.; et al. Network algorithms for detection of radiation sources. *Nucl. Instrum. Methods Phys. Res. Sect. A Accel. Spectrometers Detect. Assoc. Equip.* **2015**, *784*, 326–331. [CrossRef]

198. Wu, C.Q.; Berry, M.L.; Grieme, K.M.; Sen, S.; Rao, N.S.V.; Brooks, R.R.; Temples, C. Network detection of radiation sources using ROSD localization. In Proceedings of the 2015 IEEE Nuclear Science Symposium and Medical Imaging Conference (NSS/MIC), San Diego, CA, USA, 31 October–7 November 2015; pp. 1–2.

199. Hite, J.; Mattingly, J. Bayesian Metropolis methods for source localization in an urban environment. *Radiat. Phys. Chem.* **2019**, *155*, 271–274. [CrossRef]

200. Wu, C.Q.; Berry, M.L.; Grieme, K.M.; Sen, S.; Rao, N.S.; Brooks, R.R.; Cordone, G. Network Detection of Radiation Sources Using Localization-based Approaches. *IEEE Trans. Ind. Inform.* **2019**, *15*, 2308–2320. [CrossRef]

201. identiFINDER S900 Radionuclide Detection Systems. Available online: https://www.southernscientific.co.uk/products-by-manufacturer/flir/radiation/stride-systems#overview (accessed on 28 April 2019).

202. Passport Releases SmartShield™ v2.0. Available online: https://www.passportsystems.com/pg/products/smartshield (accessed on 28 April 2019).

MDPI

St. Alban-Anlage 66

4052 Basel

Switzerland

Tel. +41 61 683 77 34

Fax +41 61 302 89 18

www.mdpi.com

Sensors Editorial Office

E-mail: sensors@mdpi.com

www.mdpi.com/journal/sensors